STEM Education in Poverty

Brantina Chirinda • Gladys Sunzuma
Perkins Muredzi

STEM Education in Poverty

Insights from Zimbabwe and Beyond

Springer

Brantina Chirinda
Faculty of Education
University of Johannesburg
Johannesburg, South Africa

Gladys Sunzuma
Sci & Mathematics Education Department
Bundura University of Science Education
Bindura, Zimbabwe

Perkins Muredzi
Harare Institute of Technology
Belvedere, Zimbabwe

ISBN 978-3-031-92489-7 ISBN 978-3-031-92490-3 (eBook)
https://doi.org/10.1007/978-3-031-92490-3

This Springer imprint is published by the registered company Springer Nature Switzerland AG
The registered company address is: Gewerbestrasse 11, 6330 Cham, Switzerland

Preface

STEM Education in Poverty: Insights from Zimbabwe and Beyond is about Science, Technology, Engineering, and Mathematics (STEM) education in poverty and the lessons we learn from Zimbabwe. STEM education has emerged as a powerful influence in education, aiming to provide students with the skills essential for success in the twenty-first century. STEM education has gained prominence recently as societies recognise its pivotal role in driving innovation, economic growth, and global competitiveness. The world is moving towards the Fourth Industrial Revolution (4IR), where economic growth has been attributed to STEM education. STEM education is vital in this era of the 4IR, where both developed and developing countries are going through rapid change in all aspects of life. Globally, STEM education is practised differently in schools and universities. University programmes, new school curricula, instructional methods, extracurricular programmes, and teacher professional development programmes have been created to cater to STEM subjects. STEM education is envisioned to produce critical thinkers, inventors, creators, problem solvers, innovators, and professionals who will solve the world's ever-changing challenges, including inequity, food insecurity, climate change, inequality, and poverty. STEM is essential to modern education and must be noticed in impoverished areas.

Despite significant economic challenges, Zimbabwe has made remarkable strides in STEM education. The Zimbabwean government has prioritised education and has made significant initiatives in STEM education. In Zimbabwe, the evolution of STEM education has been marked by tremendous efforts to improve access, enhance instructional methodologies, and bridge the gap between education and national development. *STEM Education in Poverty: Insights from Zimbabwe and Beyond* explores the intricate relationship between STEM education and poverty alleviation in Zimbabwe, offering insights into how STEM can serve as a transformative tool for economic empowerment, innovation, and social mobility. This book provides a much-needed contribution to the international debate surrounding the optimal STEM education for students in underserved schools.

This book examines the historical, political, economic, and sociocultural factors influencing STEM education in Zimbabwe. A comprehensive analysis of

Zimbabwe's STEM education landscape, addressing important topics such as policy development, teacher perspectives, pedagogical approaches, and curriculum transformation, has been provided in this book. The book discusses the challenges that have emerged, including inadequate resources, teacher shortages, infrastructure constraints, gender disparities in STEM participation, and lack of exposure and understanding of the relevance and importance of these fields in today's world. Drawing on global best practices, it also proposes innovative strategies for enhancing STEM education in developing countries. There is a need for early exposure to STEM-based career opportunities and for more collaboration between the government, private sector, and educational institutions to address these challenges and promote STEM education in Zimbabwe.

In addition, the book focuses on the role of STEM education in breaking the cycle of poverty. STEM education can drive economic growth, foster technological advancements, and enhance social well-being by equipping students with the skills necessary for high-demand sectors. The book argues that early childhood education, community engagement, and curriculum transformation are essential in making STEM education more inclusive and impactful in Zimbabwe. Emphasis is also placed on a curriculum prioritising experiential learning, problem-solving, and critical thinking.

This book uses Zimbabwe as an example to explore STEM education in poverty, thus demonstrating that investing in STEM education is a powerful strategy for socio-economic transformation. By examining the successes and challenges of STEM education in Zimbabwe, this book offers valuable insights into how STEM education can be integrated into the curriculum in countries with developing and emerging economies. The lessons drawn from Zimbabwe's experiences can inform global efforts to expand equitable access to STEM education, ensuring that all students, regardless of their socio-economic background, can participate in the technological and scientific advancements shaping the twenty-first century.

STEM Education in Poverty: Insights from Zimbabwe and Beyond is an essential resource for policymakers, educators, researchers, and development practitioners interested in improving STEM education in underserved communities. Finally, the book contributes to ongoing debates about the future of STEM education in Zimbabwe and other developing nations.

Johannesburg, South Africa — Brantina Chirinda
Bindura, Zimbabwe — Gladys Sunzuma
Harare, Zimbabwe — Perkins Muredzi

Contents

About the Authors

Brantina Chirinda is a mathematics education researcher, teacher educator, and former public school teacher. She is currently a research associate at the University of Johannesburg and a postdoctoral researcher at the University of California, Berkeley. Dr. Chirinda holds a PhD in Mathematics Education from the University of Witwatersrand, South Africa. Her primary interest lies in the teaching and learning of mathematics among underserved populations. Using sociocultural approaches and Design-Based Research, her research focuses on (1) generating interventions that provide students with meaningful opportunities to engage with rich mathematical content and practices, particularly mathematical thinking and problem-solving, and (2) exploring ways to support teachers in underserved schools to create more equitable classrooms.

For nearly 25 years, she has taught Mathematics and Mathematics Education courses at various institutions in Southern Africa, Zimbabwe, and the United States. Dr. Chirinda is a co-author of the book *Helping Students Become Powerful Mathematics Thinkers: Case Studies and Methods on Teaching for Robust Understanding* (2023, Routledge), and editor of three volumes: *Mathematics, Science and Technology Education in Zimbabwe: Research, Policy, and Practice* (2023, Peter Lang Publishing); *Mathematical Problem-solving in South Africa: Research and Practice* (2023, Unisa Press); and *Mathematics Education in Africa: The Fourth Industrial Revolution*" (2022, Springer). She is an associate editor for the *Frontiers in Education Journal* in the STEM section and a senior editor at the *Education Journal*. She has published several conference proceedings, book chapters, and articles in accredited local and international journals. She has presented her work at national, regional, and international conferences and has been a reviewer for a wide range of local and international journals. Dr. Chirinda leads a community project, Southern African Mathematics Empowerment Network (SAMENET), envisaged to empower female mathematics teachers and learners in underserved communities.

Gladys Sunzuma is a senior lecturer in the Science and Mathematics Education Department at Bindura University of Science Education. She graduated with a BScEd in mathematics and chemistry from the University of Zimbabwe in 2002. She obtained an MScED in mathematics from Bindura University of Science Education in 2010. Dr. Sunzuma holds a PhD in mathematics education from the University of Kwa-Zulu Natal. She has served for 7 years as a high school mathematics, physical science, integrated science, and chemistry teacher in Zimbabwe, three years as a teaching assistant in the Education Department at Bindura University of Science Education, and over 12 years in her current position as a senior lecturer at Bindura University of Science Education. She has published over 36 articles in various local and international journals and has reviewed several articles from various journals. She has published one book. She is an associate editor for the *Frontiers in Education Journal* in the STEM section. Her research interests include finding ways to improve the teaching and learning of mathematics both at school and university levels and improving teacher training programmes to include better pedagogical content knowledge for teachers.

Perkins Muredzi is the Director of Academic Programmes and former Executive Dean of the School of Engineering and Technology at the Harare Institute of Technology, Harare, Zimbabwe (2005–2021). He is a global award winner of Teaching Leadership, which he received at the 2015 World Education Congress held in Mumbai, India. He sits in the Higher Education Examinations Council (HEXCO) and Zimbabwe Institution of Engineers (ZIE) boards. Eng. Muredzi taught at Vista University (SA), Harare Polytechnic, Midlands State University, and the University of Zimbabwe before joining the Harare Institute of Technology in April 2004. He graduated with a BSc and MSc in Food Processing Engineering from the Krasnodar Institute in the Russian Federation in 1991. He obtained a Master's in Intellectual Property from Africa University in 2009. He holds a PhD in Food Science & Technology from Atlantic University (USA) and is currently a PhD candidate pursuing a Doctorate in Law—Intellectual Property. He teaches Food Engineering in undergraduate, certificate, and diploma programmes. Additionally, he teaches Intellectual Property in the Master of Technology (Strategy & Innovation) and the Intellectual Property degree programme offered jointly by Africa University, the World Intellectual Property Organization (WIPO), and the African Intellectual Property Organization (ARIPO).

Eng Muredzi has published eight books, his first being one published by CBH Publishers (USA) entitled *Emerging Non-Thermal Food Processing Technologies*. Two are textbooks for "A" Level Food Science and another textbook for postgraduate students studying Food Science and Technology. Eng Muredzi serves on editorial boards for three international journals: the *Journal of Biological, Pharmaceutical and Chemical Research*; the *Journal of Leadership and Management Studies*, Department of Management & Project Management, Cape Peninsula University of Technology; and the *International Journal of Science, Technology, and Society*.

Abbreviations

2-7-4-2 Model	Two years of Early Childhood Development, Seven years of Primary Education, Four years of Secondary Education, and Two years of Higher Secondary or Vocational Training
4Cs	Creativity, Collaboration, Critical Thinking, and Communication
4IR	Fourth Industrial Revolution
A-Level	Advanced Level
AI	Artificial Intelligence
AR	Augmented Reality
ARIPO	African Intellectual Property Organization
BEAM	Basic Education Assistance Module
BEST	Better Environment Science Teaching
BSAC	British South Africa Company
BSc	Bachelor of Science
BScEd	Bachelor of Science Education
BUSE	Bindura University of Science Education
CBC	Competency-Based Curriculum
CIDA	Canadian International Development Agency
CIET	Presidential Commission of Inquiry into Education and Training
CRT	Critical Race Theory
CUT	Chinhoyi University of Technology
ECD	Early Childhood Development
ECD A	Early Childhood Development A
ECD B	Early Childhood Development B
EFA	Education for All
F1	Form 1 (Academic Track)
F2	Form 2 (Practical/Vocational Track)
GDP	Gross Domestic Product
HBE 5.0	Heritage-Based Education 5.0
HEXCO	Higher Education Examinations Council
HIT	Harare Institute of Technology
ICT	Information and Communication Technology

MHTEISTD	Ministry of Higher and Tertiary Education, Innovation, Science and Technology Development
MHTESTD	Ministry of Higher and Tertiary Education, Science and Technology Development
MoPSE	Ministry of Primary and Secondary Education
MSc	Master of Science
MScEd	Master of Science Education
MUAST	Marondera University of Agricultural Science and Technology
MSUAS	Manicaland State University of Applied Sciences
NGO	Non-governmental Organisation
NUST	National University of Science and Technology
O-Level	Ordinary Level
OECD	Organisation for Economic Co-operation and Development
PhD	Doctor of Philosophy
SAMENET	Southern African Mathematics Empowerment Network
SDC	School Development Council
SDG	Social Development Goal
SDGs	Sustainable Development Goals
STEP	Science Teaching Education Programme
STEAM	Science, Technology, Engineering, Arts, and Mathematics
STEM	Science, Technology, Engineering, and Mathematics
STI	Science, Technology and Innovation
TVET	Technical and Vocational Education and Training
UDHR	Universal Declaration of Human Rights
UN	United Nations
UNESCO	United Nations Educational, Scientific and Cultural Organization
UNICEF	United Nations International Children's Emergency Fund
USA	United States of America
UZ	University of Zimbabwe
VR	Virtual Reality
WIPO	World Intellectual Property Organization
WUA	Women's University in Africa
ZANU	Zimbabwe African National Union
ZIE	Zimbabwe Institution of Engineers
ZIMASSET	Zimbabwe Agenda for Sustainable Socio-Economic Transformation
ZIMDEF	Zimbabwe Manpower Development Fund
ZIMSEC	Zimbabwe School Examinations Council
ZJC	Zimbabwe Junior Certificate

Introduction

Abstract Education has long been recognised as a vital mechanism for fostering equality and enabling individuals to transcend socioeconomic barriers, thus facilitating sustainable development. In the context of the Fourth Industrial Revolution (4IR), the role of Science, Technology, Engineering, and Mathematics (STEM) education is paramount, as it drives innovation and cultivates essential skills for navigating the rapidly evolving technological landscape. STEM education is characterised by an interdisciplinary approach that integrates scientific and mathematical concepts with technological and engineering principles. While advanced economies harness STEM fields to propel innovation and economic growth, many developing nations, including Zimbabwe, continue to grapple with significant challenges in delivering effective STEM education, impeded by systemic poverty, inadequate infrastructure, gender disparities, and a lack of access to quality learning resources. This chapter delineates the historical context of STEM education in Zimbabwe, which dates back to the 1990s and has been influenced by reforms to enhance teacher education and practical skills in the curriculum.

Preamble

Education has long been celebrated as the great equaliser, a means through which individuals can overcome socioeconomic barriers, allowing nations to achieve sustainable development (Freire, 1970). In the Fourth Industrial Revolution (4IR) context, the significance of Science, Technology, Engineering, and Mathematics (STEM) education cannot be overstated. It is pivotal in driving innovation and fostering the critical skills necessary for navigating today's rapidly evolving technological landscape. As industries increasingly integrate advanced technologies, the demand for a workforce with a solid foundation in STEM disciplines becomes paramount. This shift enhances individual career prospects and contributes to the overall economic growth and competitiveness of nations in a globalised world. Consequently,

B. Chirinda et al., *STEM Education in Poverty*,
https://doi.org/10.1007/978-3-031-92490-3_1

prioritising and investing in STEM education is essential for equipping future generations with the tools to thrive in an increasingly complex and interconnected environment. Yet, for many developing countries, the promise of STEM education remains largely unfulfilled, impeded by systemic poverty, inadequate infrastructure, limited access to quality learning resources, lack of trained educators/teachers and gender disparities (UNESCO, 2024). In this book, the terms teacher(s) and educator(s) are used interchangeably.

STEM education encompasses technological and engineering design-based learning approaches that purposefully integrate science and mathematics concepts with technology and engineering principles (Sanders & Wells, 2006). Tsupros et al. (2009) state that STEM education is an interdisciplinary learning approach combining rigorous academic concepts with real-world applications. The interdisciplinary approach recognises that real-world challenges and innovations often require a combination of skills from multiple disciplines. By breaking down traditional subject silos, STEM education encourages students to see the connections between different fields and understand how they collectively contribute to addressing complex problems (Zhou, 2017). This interdisciplinary nature reflects the reality of modern professions, where collaboration across diverse domains is essential. Students engage with science, technology, engineering, and mathematics in ways that connect their classroom learning to their communities, workplaces, and the global economy. Thus, STEM teaching extends beyond mere knowledge acquisition to include practical, real-world applications (Madusise, 2020).

While advanced economies leverage STEM fields to drive innovation and economic growth, many developing countries struggle to provide even the most basic STEM education to their youth. This book aims to illuminate the intricate relationship between STEM education and poverty. It does this mainly by delving into the successes, challenges, and issues of the Zimbabwean educational system, which are relevant to other developing countries. The book also lays bare how access to quality STEM programmes can significantly impact economic opportunities for low-income individuals by scrutinising the barriers they face and the potential pathways to success.

Despite these challenges, numerous initiatives and case studies demonstrate that, with the right policies, investments, and community engagement, STEM education can act as a catalyst for breaking the cycle of poverty (Deshani & Hirimburegama, 2024). By equipping students with problem-solving skills, fostering critical thinking, and preparing them for high-demand careers, STEM education can establish the groundwork for economic empowerment and sustainable development (Schwab, 2016).

Ultimately, this book aims to spark conversations and inspire actionable solutions that ensure STEM education reaches the most vulnerable communities. It seeks to contribute to a future where education is not a privilege of the few but a fundamental right that empowers individuals and elevates entire societies.

STEM Education in Zimbabwe

STEM education is still a relatively young discipline, lacking a rich historical tapestry. Internationally, the focus on supporting student learning in STEM areas dates back to the 1990s when the United States National Science Foundation officially incorporated engineering and technology alongside science and mathematics in undergraduate and K-12 education. The establishment of the STEM initiative in Zimbabwe has its roots in the 1990s, too, a period marked by significant reforms introduced through the Nziramasanga Commission and the Zimbabwe Agenda for Sustainable Socio-Economic Transformation (ZIMASSET), which was coined the Zimbabwe economic blueprint. The Nziramasanga Commission was tasked with evaluating and recommending improvements to the educational system. Its key focus areas included enhancing teacher education, promoting science and technology, and developing practical skills in the curriculum. This period marked a significant emphasis on enhancing education in STEM fields to foster economic development and innovation within Zimbabwe.

In 2012, the Zimbabwean government introduced the "Second Science, Technology, and Innovation Policy" framework. This initiative was designed to enhance education and skills development among students, focusing on equipping them with advanced competencies. The primary goal was to prepare students for active participation in local and global economies (Dube, 2018). The Ministry of Higher and Tertiary Education has emphasised the strategic economic importance of STEM subjects, prioritising them (Ngwenya et al., 2017). To underscore this, in 2016, a STEM program was implemented, which provided free education to all advanced-level (A-level) students enrolled in STEM subjects. The objectives of this initiative were to increase the number of STEM students enrolling in STEM degree programs at the country's universities by 2018, stimulate interest in Mathematics, Biology, Physics, and Chemistry as foundational pillars of STEM competencies and promote STEM careers in alignment with ZimAsset's value-addition and beneficiation thrust, to train and develop cutting-edge skills to support Zimbabwe's industrialisation goals and enhance the country's global competitiveness (Ngwenya et al., 2017). Dube (2018) further notes that this program was designed to address the growing demand for expertise in STEM areas and minimise negative attitudes towards Mathematics and Science subjects.

Despite its positive intentions, the initiative faced significant limitations. The major limitation was that STEM initiatives are more effective when introduced to students during early childhood development and primary school rather than later at A-level (Pakombwele & Tsakeni, 2024). The other weakness of the program was that financial support was offered exclusively to students at the A-level. As a result, numerous individuals who qualified for higher education after completing their A-levels in STEM could not afford university costs. The focus on A-level support restricted access to university for a broader range of students, highlighting a critical gap in the initiative's overall effectiveness in promoting equitable educational opportunities. Besides financial support for students, Zimbabwe still faced other factors hindering STEM education, such as inadequate scientific infrastructure at institutions, a lack of necessary teaching resources, insufficient subject matter knowledge by educators, curriculum deficiencies, and educator shortages (Dekeza & Kufakunesu, 2017; Kadziya & Ndebele, 2020).

Recruiting and retaining qualified STEM educators has proved a big challenge for the Zimbabwean government due to a shortage of subject knowledge experts. STEM experts often opt for more lucrative opportunities away from the teaching profession. This deficit in qualified educators leads to the employment of inadequately prepared and unqualified educators, impacting the provision of quality STEM education to students (Kadziya & Ndebele, 2020). Without qualified STEM educators, students may miss vital concepts and hands-on experiences that foster their ability to tackle future intricate challenges. In addition, it is common for students to find it challenging to cultivate essential analytical and complex problem-solving skills, which are increasingly critical for success in various 4IR careers, ranging from engineering and data science to healthcare and environmental technology. The increasing demand for skilled STEM educators has prompted significant educational reform initiatives by the Zimbabwean government. In 2018, the government responded to this urgent need by launching a comprehensive professional development initiative focused on STEM education, the Science Teaching Education Programme (STEP) (Tshili, 2023). The STEP initiative explicitly targets education colleges, aiming to enhance the efficacy and quality of STEM instruction through rigorous training for secondary school science teachers. Its primary purpose is to produce qualified STEM teachers for Zimbabwean secondary schools.

We conducted a document analysis of the STEP and found that the programme fundamentally lacks opportunities for student teachers to innovate, experiment, and engage in critical thinking. This deficiency undermines essential elements of effective STEM education, resulting in insufficient training for teachers. Consequently, the STEP programme requires substantial changes and enhancements to equip student teachers with the vital skills and competencies needed to deliver high-quality STEM education in secondary schools across Zimbabwe (Mufanechiya & Makgalwa, 2024).

Through a comprehensive analysis, this book, authored by us leading scholars in STEM education in Zimbabwe, highlights the importance of equitable educational frameworks and how policy choices influence access and effectiveness of STEM education for underserved populations. Our personal and professional backgrounds, educational experiences, and research perspectives have profoundly shaped the content of this book. Next, we provide a brief overview of our personal and academic journeys to enhance your understanding of the rationale behind the structure and themes of this work. Drawing from our insights, we aim to clarify how our experiences have informed the concepts presented within the book.

Authors' Positionality

Dr. Brantina Chirinda

I embody a complex intersection of diverse identities, each layer intricately shaped by my unique lived experiences. These experiences—encompassing cultural background, professional and personal challenges, and significant moments—serve as lenses through which I view and interpret the world around me. This multifaceted identity informs my understanding of various social dynamics and profoundly influences the themes and narratives explored in this book. As I share my insights and reflections, I aim to convey how these intersecting identities have enriched my perspective and contributed to my growth as a STEM education researcher, teacher educator, and former public school teacher.

I grew up in Zimbabwe, in a disadvantaged village called Buhera, a rural area away from the world's vast technological advances. From learning to count using stones under trees in our rural area to studying mathematics, statistics, and physics for my Bachelor of Science degree, I have found STEM education fascinating due to its prominent application in our lives. After earning a bachelor's degree, I taught mathematics at a secondary school in contexts of disadvantage in Zimbabwe. This was when I recognised firsthand the barriers that socioeconomic challenges pose to accessing quality education. This experience instilled in me a strong sense of responsibility to advocate for equitable educational opportunities, particularly in STEM fields, which are critical for fostering innovation and economic growth.

When I started teaching at a secondary school, I had no adequate training in teaching. Not knowing there were alternatives, I fell back on the teacher-centred model, which was how my professors at university had taught me. Unfortunately, no one ever taught them how to teach either, and as a new teacher, I was the only source of information in my classes, and my students were passive receivers of knowledge. Nonetheless, my students forgot a large chunk of the mathematics they learned after an exam. After a few years of teaching, I began to realise that my students forgot the mathematics because they did not understand the concepts deeply. The intense focus on rote memorization and exam preparation often led to a shallow understanding of the material, causing important mathematics concepts to slip from memory shortly after the assessments were completed. This phenomenon highlighted the need for me to embark on graduate studies to explore more effective learning strategies that encourage long-term retention and more profound comprehension rather than temporary performance for the sake of passing exams.

My graduate studies involved exploring the teaching and learning of mathematics in underserved populations. My research primarily centred on how educational practices can be shaped to support students from disadvantaged backgrounds in learning mathematics. This focus gave me valuable insights into the systemic hurdles students face in low-income environments and the urgent need for targeted interventions in STEM education.

As an educator and researcher, I strive to bridge the gap between theoretical concepts and the lived experiences of my students. I am committed to addressing educational disparities and promoting equitable access to STEM education in underserved communities. STEM is a universal language that transcends borders and empowers individuals to navigate complex challenges. Yet, for too long, underserved students have been denied equitable access to the transformative power of STEM education. My research agenda is rooted in the belief that every student, regardless of their background, deserves the opportunity to engage in meaningful exploration of STEM concepts. My work involves developing interventions that prioritize inclusivity. Using Designed Based Research and sociocultural approaches, my current research focuses on (1) generating interventions that assist in providing students with meaningful opportunities to engage with rich mathematical content and practices, particularly mathematical thinking and problem solving, and (2) exploring and understanding ways to support STEM teachers in underserved schools to make their classrooms more equitable.

Throughout my teaching journey, I've had the privilege of guiding students in a diverse array of classroom environments in Zimbabwe, South Africa and the United States of America. Each setting has offered unique challenges and inspiring moments, enriching my experience and approach to STEM education.

In particular, as a STEM teacher educator, I am informed by Critical Race Theory (CRT) in my commitment to creating an inclusive and equitable environment where every student can thrive. CRT provides a critical race-based lens to view the historical consequences that segregation has had on the STEM education of non-whites. Drawing upon the work of Matsuda et al. (1993), the central tenets of CRT are to recognize that racism is endemic, and to reject meritocracy and objectivity because racism advantages and disadvantages individuals based on race. According to Delgado and Stefancic (2012), the primary goal of CRT is to study and potentially transform the relationships among racism, race, and power. More broadly, CRT teaches us that all historically marginalized groups have been systemically disadvantaged and that efforts must be undertaken to undo these disadvantages. For this reason, I have devoted most of my career to undoing the unpleasant impacts of structural racism by supporting STEM students who have been marginalized. I am sensitized to the needs of groups that have been "othered" or marginalized and apply my experiences to every population.

My commitment to fostering social change through education is also reflected in my collaborations with local organizations and educational institutions aimed at improving access to quality STEM education for underprivileged youth. Working with these communities has deepened my understanding of the unique strengths and abilities of students from diverse backgrounds, and I emphasise the importance of culturally relevant pedagogy in my writings.

In summary, my positionality is shaped by my personal experiences, academic background, and a steadfast commitment to advocating for equitable STEM education. I acknowledge the multifaceted nature of poverty and its impact on education, and I invite others to consider how inclusive teaching practices can empower all students to succeed in STEM fields. Through my work, I highlight STEM education as a transformative tool for addressing societal inequities and fostering a more just and equitable future.

Dr. Gladys Sunzuma

I bring a unique positionality to my work, shaped by a blend of personal experiences, cultural background, and scholarly endeavours. Born and raised in Zimbabwe, I navigated a diverse tapestry of cultures, fostering a deep appreciation for the intricacies of human experiences. As a seasoned educator and researcher with a multifaceted background rooted in the sciences and education, my academic and practical experiences significantly shape my position as an author in the realm of mathematics education and STEM. My foundational journey, marked by a first degree in Chemistry and Mathematics, laid the groundwork for a comprehensive understanding of the sciences. This dual expertise reflects a strong analytical foundation and provides practical insight into the interdisciplinary nature of STEM fields. My early academic trajectory demonstrates a commitment to theoretical principles and real-world applications. Having pursued my education in chemistry and mathematics, I developed a keen interest in mathematics pedagogy, STEM education, and teacher development. This academic journey not only equipped me with theoretical frameworks but also instilled a commitment to critically examining the intersections of STEM subjects and their broader societal implications.

Building upon this foundation, I pursued a second degree in Mathematics Education, marking a deliberate shift towards pedagogy and instructional methodologies. My subsequent Ph.D. in Mathematics Education further solidified my dedication to advancing teaching practices within the field. This academic trajectory underscores my commitment to bridging the gap between content expertise and effective pedagogy. The wealth of experience gained from teaching mathematics and chemistry in Zimbabwean high schools, coupled with leadership responsibilities as the head of the science and mathematics department, provides me with a unique perspective on the challenges and opportunities within STEM education at the secondary level. This practical engagement with Zimbabwe's educational landscape enriches my writing with a grounded understanding of the local context.

As a senior lecturer at Bindura University of Science Education, I continue to contribute significantly to the field. My role involves teaching mathematics content and pedagogy courses while engaging in research. With an extensive publication record comprising articles and book chapters, I am committed to advancing knowledge in mathematics education. My research focus on pedagogy and teacher education in mathematics underscores my dedication to enhancing the quality of mathematics instruction. Additionally, my role as a reviewer for various academic journals reflects my active participation in shaping discourse within the scholarly community.

As one of the authors of this book, I am exploring the intersection of STEM and poverty in Zimbabwe, further extending my reach beyond academic circles. This project reflects my commitment to addressing broader societal challenges through education and showcases my dedication to fostering positive change through scholarly contributions. My STEM worldview informs this work by emphasizing how science, technology, engineering, and mathematics can be leveraged to mitigate poverty and promote sustainable development in Zimbabwe. By exploring the socio-economic dimensions of STEM education, I aim to illuminate pathways for empowering marginalized communities through knowledge and innovation.

Dr. Perkins Muredzi

I grew up in a society that yearned for access to quality education before the independence of the Republic of Zimbabwe. I received primary education that promoted the British education system and went on to enrol for secondary and senior school just as the Zimbabwean education system was transforming. The British Cambridge school's ordinary and advanced certificate programmes transformed, and the Zimbabwe Schools Examination Council was born and established.

My worldwide view of STEM education revolves around an understanding that there is a growing global recognition of the importance of STEM education in addressing societal issues, promoting innovation and driving economic growth.

Through experience, I learnt a lot regarding the new education system and witnessed the education transformation championed by the new independent government. I vividly remember the choice of embarking on further studies in STEM subjects and my eventual passion and enrolment into undergraduate and postgraduate studies in engineering. I received tertiary education in the Russian Federation and the USA and hence have an understanding of trending education contexts of the East and West.

This book contribution serves to champion the promotion of STEM studies cognisant of the prevailing poverty in most developing countries, using my country, Zimbabwe, as an example. Addressing the impact of poverty on STEM education requires targeted interventions that provide equitable access to resources, quality infrastructure, qualified teachers and opportunities for involvement in STEM-related activities while fostering inclusivity and support for underprivileged students. By integrating science, technology, engineering and mathematics into the curriculum, providing financial support and incentives, improving the quality of education and fostering collaboration between academia and industry, Zimbabwe is laying the foundation for a generation of STEM professionals who can contribute to the country's economic growth and global competitiveness. Herein lies the lessons and takeaways from this publication for educators worldwide.

I have played a role in establishing Midlands State University and Harare Institute of Technology and have participated in curriculum development in the two institutions that culminated in the conferment of a global award in Education Leadership at the World Education Congress held in Mumbai, India, in 2015. I have held the position of executive dean for Engineering & Technology and Industrial Sciences and have championed STEM education and the Education 5.0 Doctrine adopted by the government of Zimbabwe.

About This book

The first chapter discusses the development and challenges of STEM education in Zimbabwe since the 1990s. The second chapter focuses on Zimbabwe's STEM education model. It provides a brief overview of the country's education context and a nuanced understanding of the historical, political, economic, and sociocultural factors that have shaped the trajectory of the Zimbabwean education system. It identifies key achievements, challenges, and future directions for policymakers, researchers, and practitioners, offering valuable insights for anyone interested in understanding the dynamics of STEM education in Zimbabwe and its potential for improving people's lives and advancing Zimbabwe's overall development.

Chapter "Knowledge, Curriculum and STEM Education: Theoretical Underpinnings" discusses the critical role of STEM education in Zimbabwe's educational system and its alignment with the country's goal to integrate into the global economy. It highlights how the Ministry of Higher and Tertiary Education, Innovation, Science, and Technology Development is implementing Education 5.0 to foster innovation and economic growth. STEM education is portrayed as fundamental for preparing students to meet the challenges of a rapidly changing world, emphasising experiential learning, problem-solving, and critical thinking. The chapter discusses the philosophical roots of STEM education, which can be traced back to the early twentieth-century progressive education movement, notably influenced by thinkers like John Dewey, who advocated for hands-on learning and student engagement. In Zimbabwe, STEM is framed as a constructivist approach where students actively participate in learning, connecting theoretical concepts with practical experiences. This student-centered philosophy also encourages the holistic development of students, fostering critical thinking, creativity, and problem-solving skills. Ultimately, the text underscores the importance of STEM education in cultivating well-rounded individuals ready to contribute positively to society.

Chapter "STEM Education Pedagogy in Developing Country Contexts: Relevance and Challenges" highlights the relevance and practices of STEM education pedagogy in developing country contexts, focusing on Zimbabwe as a case study. The need for STEM education is driven by economic diversification and the shift towards knowledge-based growth, emphasizing the importance of the four Cs of twenty-first-century education: Creativity, Collaboration, Critical Thinking, and Communication. The chapter highlights the critical role of an inquiry-based, adaptable framework in nurturing a love of learning among students. Despite the growing demand for STEM professionals, developing countries face challenges such as resource constraints, outdated infrastructure, and cultural barriers that hinder the development of a qualified workforce. A survey of 798 educators in Zimbabwe revealed a reliance on traditional teaching methods and disparities in STEM education practices between rural and urban areas.

Chapter "Zimbabwe STEM Education: Policies and Practices" uses document analysis to examine Zimbabwe's STEM education policies and practices. It unpacks the Zimbabwean government's initiatives, curriculum integration, teacher development, infrastructure, partnerships, and inclusivity efforts, highlighting challenges in implementing effective STEM education and suggesting strategies to overcome them. The chapter also evaluates the impact of STEM education on students' performance, career opportunities, and national development and concludes by giving recommendations regarding measures to advance STEM education in Zimbabwe. Chapter "Access and Equity in STEM Education" provides a comprehensive overview of STEM education's access and equity issues in Zimbabwe and emphasises the importance of creating equal opportunities for all students to engage and succeed in STEM fields. In Zimbabwe, the emphasis is on improving STEM education by increasing practical learning and shifting from a content-based curriculum to a competency-based one. The chapter notes that underrepresented groups, like females and rural students, still face significant barriers to accessing STEM education in Zimbabwe. Despite efforts, disparities in access, quality, and resources persist in STEM education in Zimbabwe.

Chapter "Teachers' perspectives on the Effectiveness of STEM Education" highlights the critical significance of teachers' perspectives regarding STEM education and elucidates the various benefits of STEM education and the challenges encountered during its implementation. By considering teachers' perspectives and addressing the challenges they face, the chapter aims to contribute to improving STEM education and ultimately better preparing students for future success in STEM fields. The findings from the empirical investigation show that Zimbabwean teachers perceive STEM education as essential for cultivating vital skills among students. Community involvement is a key factor in enhancing student engagement and motivation. The availability of resources is deemed crucial for the effective delivery of STEM education. The findings ultimately aim to assist educational institutions globally in designing teacher empowerment courses, planning future education policies, and understanding teachers' needs to improve STEM education.

Chapter "STEM Education as a Praxis to End Poverty: A Curriculum Turn" dwells on how STEM education can end poverty through curriculum transformation. STEM education possesses the potential to disrupt the cycle of poverty since it effectively prepares students for employment in high-demand sectors. In Zimbabwe, numerous underprivileged communities encounter substantial obstacles in accessing quality STEM education. Initiatives such as the Basic Education Assistance Module (BEAM) aim to support these communities. Chapter "STEM Education as a Praxis to End Poverty: A Curriculum Turn" proposes that a transformative curriculum emphasises the practical application of STEM skills. The chapter notes that Early Childhood Education is paramount in cultivating interest in STEM subjects. Community-centric content enhances the relevance of educational experiences, and the incorporation of global perspectives aids students in comprehending poverty as a universal issue.

Chapter "The Future of STEM Education in a Developing Country: Re-aligning Approaches for Successful Implementation" argues that the future of STEM education in developing countries depends on realigning approaches that prioritise local contexts and resources. Developing countries can create a supportive and inclusive environment that prepares students for successful careers in STEM fields by building partnerships, creating relevant curricula and providing ongoing teacher training. The chapter advocates that STEM education is crucial for fostering sustainable development within developing nations and addresses global challenges by cultivating a skilled workforce equipped to navigate complex issues. The chapter argues that it is essential to modernise STEM curricula and integrate advanced technology into educational frameworks. Providing adequate resources and facilities is vital for facilitating practical learning experiences. Efforts must be directed towards ensuring equal access and support for underrepresented groups in STEM fields. Collaborative approaches should be fostered to improve educational outcomes and address existing disparities in access to education. By doing so, every student can engage with scientific concepts more holistically and meaningfully and develop critical thinking skills, ultimately preparing them for successful careers in STEM fields.

Chapter "Looking Ahead" concludes the book by offering key insights into how Zimbabwe has made significant progress towards providing high-quality education in STEM despite limited resources. The chapter offers a path to leverage STEM education for poverty eradication by giving general strategies to enhance it in Zimbabwe. The chapter provides recommendations for policymakers, educators, and activists in the education sector, especially in underserved, underdeveloped, and developing countries worldwide.

References

Dekeza, C., & Kufakunesu, M. (2017). Implementation of STEM curriculum in rural secondary schools in Zimbabwe: Limits and possibilities. *Journal of Emerging Trends in Educational Research and Policy Studies, 8*(1), 11–15.

Delgado, R., & Stefancic, J. (2012). *Critical race theory: An introduction* (2nd ed.). New York University Press.

Deshani, N. A. D. K., & Hirimburegama, K. (2024). Impact of STEM education in overcoming poverty. *KDU Journal of Multidisciplinary Studies, 6*(2), 67–78.

Dube, P. S. (2018). An investigation of the Science Technology Engineering and Mathematics (STEM) initiatives in the Zimbabwean education system. *International Journal for Cross-Disciplinary Subjects in Education (IJCDSE), 9*(1), 3329–3334.

Freire, P. (1970). *Pedagogy of the oppressed*. Seabury Press.

Kadziya, L., & Ndebele, C. (2020). Towards improving the quality of teaching and learning of STEM subjects in rural post-colonial Zimbabwe. *African Renaissance, 17*(4), 141–165.

Madusise, S. (2020). Possibilities for implementation of STEM education in Zimbabwean under-resourced mathematics classrooms. *Journal of African Interdisciplinary Studies, 3*(7), 110–124.

Matsuda, M. J., Lawrence, C. R., III, Delgado, R., & Crenshaw, K. W. (Eds.). (1993). *Words that wound: Critical race theory, assaultive speech and the first amendment*. Westview Press.

Mufanechiya, A., & Makgalwa, M. M. (2024). Re-visioning Zimbabwe's Science, Technology, Engineering and Mathematics for secondary schools through the science teacher education programme. *Journal of Education and Learning Technology, 5*(6), 111–129.

Ngwenya, B., Pelser, T., & Sibanda, R. (2017). Success as the knowledge economy the success or failure of the Zimbabwe's STEM initiative. In *International conference on intellectual capital and knowledge management and organisational learning* (pp. 169–178). Academic Conferences International.

Pakombwele, A., & Tsakeni, M. (2024). Designing learning centres that facilitate knowledge construction on science process skills in early childhood classrooms. *Journal of Ecohumanism, 3*(7), 755–764. https://doi.org/10.62754/joe.v3i7.4241

Sanders, M., & Wells, J. G. (2006). *Integrative STEM education*. Retrieved from http://www.soe.vt.edu/istemed/

Schwab, K. (2016). *The fourth industrial revolution*. World Economic Forum.

Tshili, N. (2023, August 29). STEM training programme expanded. *The Herald: Zim Papers*. Retrieved January 12, 2024, from https://www.herald.co.zw/stem-training-programme-expanded

Tsupros, N., Kohler, R., & Hallinen, J. (2009). *STEM education: A project to identify the missing components*. Intermediate Unit 1: Center for STEM Education and Leonard Gelfand Center for Service Learning and Outreach, Carnegie Mellon University.

UNESCO. (2024). *Towards 2030 and beyond: Challenges and opportunities for education transformation*. UNESCO International Bureau of Education.

Zhou, M. (2017). *STEM implementation in rural secondary schools: A case of Kwekwe Zhombe district*. Undergraduate research project. Midlands State University.

The Education Context of Zimbabwe

Abstract The chapter provides a comprehensive overview of Zimbabwe's educational context, offering a nuanced understanding of the historical, political, economic, and sociocultural factors that have shaped the trajectory of the Zimbabwean education system. It highlights key achievements, challenges, and future directions for policymakers, researchers, and practitioners. It delivers valuable insights for anyone interested in the dynamics of Science and STEM education in Zimbabwe and its potential to improve lives and advance the nation's overall development.

Historical Overview

Colonial Government in Zimbabwe Until 1980

The British South Africa Company (BSAC) arrived in Rhodesia in the 1890s. Rhodesia is now represented by the countries Zimbabwe, Malawi, and Zambia. The Company administration of Rhodesia created Christian missionary schools to serve local communities. Missionary schools provided an education for the indigenous population that focused on agricultural production and industrial development, including carpentry and building. The Company administration of Rhodesia limited education and censored knowledge in schools for local communities and reinforced the superiority of White settlers, even though they were the minority of the population (Atkinson, 1972). Missionary schools perpetuated social and economic repression of the indigenous population by reducing their chances of earning well-paying jobs or positions of power (Kanyongo, 2005). Education to Africans offered limited academic and foundational skills to promote labour exploitation and indentured servitude (Mapako & Mareva, 2013). Restricted access to a quality education kept Africans subordinate and inferior, to White colonists to advance British political and economic gains.

Due to significant investments in education by the Southern Rhodesian government prior to 1960, Europeans received disproportionately more funding for

B. Chirinda et al., *STEM Education in Poverty*,
https://doi.org/10.1007/978-3-031-92490-3_2

education than the majority African population. This disparity arose because the Rhodesian government controlled access to quality schools based on race, prejudice, and ethnicity. The segregation of schools in terms of funding was most extreme in the 1970s; despite Europeans comprising only a small percentage of Zimbabwe's population, they were allocated around 90% of government spending on education. Funding for secondary schools was also disproportionately offered to Europeans instead of Africans. In the 1970s, only 43.5% of African children attended school, and only 3.9% of these children enrolled in secondary school, reflecting the enrollment imbalances of that decade.

In 1979, the new government of Zimbabwe Rhodesia called for an educational reform that established a three-tier school system. The Education Act of 1979 regulated access to each type of school through a zoning system based on residency (Shizha & Kariwo, 2011). Prior to the act, Zimbabwe's education system was divided into African and European schools. After the shift in policy and leadership, the education system was divided into government schools, community schools, and private schools. Government schools were further divided into three categories known as Group A, B, and C. White students historically attended Group A schools, which provided highly trained teachers and a quality education. These schools were situated in white suburbs that denied housing opportunities for Africans, thereby reinforcing segregation based on ethnicity and race. Group B schools required a low fee, while C schools did not charge any fee beyond educational materials. Both were exclusively available for African students. Group B and C schools had fewer resources, less funding, and less qualified faculty compared to Group A schools.

During the colonial era, missionary organizations and limited government initiatives expanded access to primary education more than secondary education (Hungwe, 1994). At the secondary level, the curriculum was heavily influenced by the British academic system, with the most prestigious qualification being the Ordinary Level (O-Level) certificate. This 2-year program, typically undertaken by students aged 14–15, was administered by the University of Cambridge Local Examination Syndicate. The O-Level certificate was internationally recognized and provided pathways to careers in teaching, nursing, higher education, and skilled trades.

Mwase and Marovah (2023) explain that during this period, the education system in Zimbabwe was divided into F1 and F2 categories. The F1 curriculum for the whites focused on academic subjects such as Geography, Mathematics, Science, History, and English. In contrast, the F2 curriculum for black students emphasized practical subjects like Agriculture, Home Economics, Building, and Woodwork. However, Mwase and Marovah (2023) highlights some shortcomings of the F2 system, stating that it produced only manual laborers without management skills. As a result, graduates from the F2 system often lacked the ability to conduct cost assessments for projects or interpret data, limiting their capacity to take on roles beyond simply following instructions. Within the F1 system, English language, science, and mathematics were particularly significant, as passing these subjects increased one's prospects of entering well-paid technical or scientific fields (Hungwe, 1994). For the F1 level, there was an Internal Junior Certificate Syllabus which was divided into separate sections for Physics, Chemistry, and Biology for convenience. At the O-level, depending on the quality of science teaching and the available resources at individual schools, students could either take the Cambridge O-level General Science or a combination of Physical Science and Biology. The latter combination was typically offered to students in schools with more advanced science facilities and higher-quality teaching staff. Science and mathematics have long been associated with technological progress, economic development, and national growth.

National Education Reform in 1980

The *Chimurenga* War from 1964 to 1979, a 15-year guerrilla war, catalysed the shift in power from British colonial rule to de jure sovereignty in 1980 (Andor et al., 1999). The Zimbabwe African National Union (ZANU) party won the national election in 1980 and democratised education by offering free and compulsory primary and secondary education to all children in Zimbabwe. The party's claims were backed by the national constitution, which recognises education as a basic human right. All primary school tuition fees were abolished after independence (MacKenzie, 1988). Dr. Dzingai Mutumbuka was elected the Minister of Education to support Zimbabwe through another education reform and to keep students in school. His leadership changed the climate of the education system because the Ministry of Education focused on fostering self-sufficient students who are productive, motivated and dedicated citizens. The government allocated 17.3% of the national budget toward education, which was politically considered an education miracle (Mackenzie, 1988). Ultimately, Zimbabwe's education system reform ensured equal access to education by providing primary and secondary education to all children (Matereke, 2012).

1980s and 1990s

Since independence, the government has focused on providing equal and free education for all through the rapid expansion of educational resources to keep up with the demand (MacKenzie, 1988). Within 1 year, the education system nearly doubled the number of students it served from 885,801 students to 1,310,315 students in primary and secondary education. Exponential increases in the number of students attending school heightened the need for more infrastructure and teachers (Brooks World Poverty Institute, 2009).

The modern scientific and technological revolution emerged in some parts of the world, leading to rapid development. (Hungwe, 1994). It was expected that Zimbabwe, after acquiring sovereign status, should affirm a faith in science and technology as the tools for accelerating development and enhancing state power and prestige.

Opening an educational planning conference in 1982, Mugabe argued for a 'greater emphasis on scientific thought and technological skills' if education was to 'bear the fruit that we expected of it' (Press Statement 1982). Science was, therefore, to be given a prominent place in education, and steps were to be taken to make it sensitive to local conditions.

Within the new curriculum, the Zimbabwe Junior Certificate Science Syllabus was introduced in 1982. The syllabus was designed to offer a foundational course in general science and serve as an appropriate preparation for the O-level syllabus. In 1987, a new O-level Science Syllabus was introduced. The syllabus focuses on the practical study of science and technology applications in Zimbabwe, aiming to extract scientific concepts from real-world practices. It was divided into five main sections: science in agriculture, science in industry, science in energy use, science in structures and mechanical systems, and science in the community. Within this curriculum, three subjects were particularly valuable: English language, science, and mathematics. Achieving passes in these subjects significantly improved one's chances of gaining training in the modern sector, particularly in well-paid technical or scientific fields.

Teachers were in high demand immediately following Zimbabwe's independence. In the mid-1980s, thousands of refugee children from Mozambique immigrated to Zimbabwe, increasing the number of children attending public schools and the demand for teachers. The Minister of Education brought in teachers from Australia, Britain, and Canada for a short period to fill the teaching gaps. Schools expanded their human resources to serve as many children as possible with limited infrastructure by practicing "hot-seating," also known as double session schooling. Hot-seating is the practice of offering class in the morning to half the students and in the afternoon to the other half (Chikoko, 2007). Hot-seating was still insufficient to meet the population's demands; therefore, the Ministry of Education expanded teacher education colleges rapidly by providing on-the-spot teacher training. In 1986, 8000 additional teachers were trained to meet national demands.

Communities also rapidly built more infrastructure for education. For example, from 1979 to 1984, the number of primary schools in operation increased by 73.3%, and the number of secondary schools increased by 537.8%. Despite the challenges of educating students on a large scale, Zimbabwe claimed to have achieved universal primary education by the end of the 1980s. By the 1990s, primary schooling was nearly universal, and over half the population had completed secondary education (Brooks World Poverty Institute, 2009).

UNICEF claims that the country's education system was once the most developed on the continent. However, the system continues to suffer from a contemporary decline in public funding linked to hyperinflation and economic mismanagement. A 40% decrease in GDP from 2000 to 2008 marked a period of economic downturn in the first decade of the twenty-first century. Social expenditures on health and education also decreased by more than half.

By the end of 2008, most schools and hospitals were shut down due to thousands of teachers leaving the profession, an economic crisis, an increase in HIV and AIDS, and an outbreak of cholera in 2008, leading to a national epidemic (Chikoko, 2007).

UNICEF (2011) reported that 94% of rural schools, serving most of the population, closed in 2009. During this period, UNICEF (2011) also claimed that attendance rates plummeted from over 80% to 20% (UNICEF, 2011). The economy regained momentum after 2009 once the Government of National Unity was formed. This was an inclusionary government developed to resolve national challenges. The Government of National Unity suspended the Zimbabwean currency to implement full dollarisation, reducing hyperinflation and increasing social expenditure (Chikoko, 2007).

Zimbabwe's focus on expanding education opportunities for the past 25 years has led to national accomplishments, including achieving a literacy rate comparable to other African countries at 51% among aged 15–24. As of 2014, 3,120,000 students were enrolled in primary and secondary education and 76% of these students were enrolled in primary education (Education Policy and Data Center, 2014).

There has been notable progress concerning access to education in Zimbabwe. According to Dokora (2015), enrolment figures increased at primary and secondary levels from 1.6 million at independence in 1980 to 4.1 million in 2015. However, Moyo (2015) reports that 0.03% of the population in Zimbabwe has acquired some STEM education at the tertiary level. This scenario has been viewed as one of the inhibiting factors of progress with respect to unlocking economic development in Zimbabwe through value addition, beneficiation, and industrialisation.

Zimbabwe Education Structure

Education underpins the social well-being of a country's citizens. Therefore, education systems need to be carefully planned to serve society's economic and social objectives, such as inclusive growth and development, the pursuit of social justice, and greater access to knowledge. A country's education context provides insights into its development, challenges, and aspirations, and Zimbabwe is no exception. In Zimbabwe, education has played a critical role in shaping the nation's identity, promoting social mobility, and driving economic growth.

Education in Zimbabwe falls under the jurisdiction of the Ministry of Primary and Secondary Education for primary and secondary education and the Ministry of Higher and Tertiary Education, Innovation, Science, and Technology Development for higher education. The Cabinet of Zimbabwe regulates both ministries. The education system in Zimbabwe encompasses 13 years of primary and secondary schooling, which is known as the 2-7-4-2 model. This model divides schooling into 2 years of early childhood development (ECD), 7 years of primary education, 4 years of secondary education, and 2 years of higher secondary or vocational training. This model aims to provide a basic educational foundation for all children while allowing them to specialise in specific fields during higher secondary education. This educational stratification enables students to acquire targeted skills and knowledge relevant to their future careers.

The Zimbabwean school year runs from January to December. The school year consists of 40 weeks, divided into three terms, with a month break between each term. The education system can be split into formal and non-formal education. Formal education refers to the structured process wherein students attend school and progress through all stages of the educational cycle. Education is institutionalised, intentional, and planned through public organisations and recognised private entities. Formal education systems are thus acknowledged as such by the Ministry of Primary and Secondary Education and the Ministry of Higher and Tertiary Education, Innovation, Science, and Technology Development.

Primary Education

Primary education in Zimbabwe spans 9 years, encompassing both infant and junior education. Infant education consists of 4 years of schooling, covering Early Childhood Development A (ECD A), Early Childhood Development B (ECD B), Grade 1, and Grade 2. The national curriculum for ECD is designed to support the holistic development of children encompassing physical, cognitive/intellectual, creative, social, moral, emotional, and health. The six core curriculum areas include an outdoor playground, block area, drama/house area, art and craft area, music and movement area, book area, and science/discovery area. Following this, junior education comprises a 5-year learning journey from Grade 3 to Grade 7. After Grade 7, students undergo examinations that signify the completion of the primary education cycle. The primary school curriculum in Zimbabwe provides a well-rounded education, integrating academic, practical, and life skills. Core subjects include Mathematics, English, Indigenous Languages, and Science and Technology. Humanities cover Social Studies, Heritage Studies, and Religious and Moral Education, while practical subjects focus on Agriculture, Information and Communication Technology (ICT), Visual and Performing Arts, and Physical Education. Life skills subjects such as Guidance and Counselling and Family and Heritage Studies promote personal development. The primary school curriculum, which includes Mathematics, Science and Technology, and ICT, lays the foundation for STEM learning by equipping students with problem-solving, analytical, and digital skills from an early age.

Secondary Education

The secondary education system in Zimbabwe spans 6 years and is divided into lower and upper secondary education. The lower secondary level comprises a 4-year cycle (Forms 1–4), culminating in the awarding of the Ordinary Level Certificate. Upon completing these 4 years, students can enter the labor force, enroll in tertiary education, or continue their secondary education for an additional 2 years in upper secondary. At the upper secondary level, students engage in advanced studies designed to prepare them for tertiary and university education, equipping them with the knowledge and skills necessary for higher learning and professional opportunities.

The secondary school curriculum in Zimbabwe is structured to provide students with a broad academic foundation while preparing them for higher education and careers. It includes core subjects such as Mathematics, English, Indigenous Languages, and Science, alongside Humanities, Commercial, Technical, and Vocational subjects. STEM education is a key component of the curriculum, fostering innovation, problem-solving, and critical thinking skills. Mathematics, Physics, Chemistry, Biology, and Computer Science form the backbone of STEM education, while Technical and Vocational subjects, including Engineering Graphics and Design, Agriculture, and ICT, provide practical applications.

Tertiary Education

Various institutions offer Tertiary education in Zimbabwe, including agricultural colleges, teachers' colleges, polytechnics, and universities. Technical and Vocational Education and Training (TVET) colleges focus on equipping students with practical skills directly applicable in industry while also providing business studies and social sciences courses. Industry plays a crucial role in facilitating apprenticeships for students. Teachers' colleges train primary and secondary school teachers, requiring candidates to have at least five O-level passes at grade C or better for admission into diploma programs. Conversely, universities provide pathways to certificates, diplomas, and degrees, with preference given to students who have satisfactory A-level passes. There is also an avenue for those with relevant qualifications from other colleges, including teachers' and technical colleges, to gain admission. Successful undergraduate students have the opportunity to progress to postgraduate studies, and those who excel in their master's programs can pursue doctorate degrees. Each stage of the structure has objectives that must be met; however, each level ultimately prepares a person for the levels that follow.

The Zimbabwean education system can be divided into formal and non-formal education. On the other hand, informal education refers to part-time schooling for students. Students do academic and professional courses. The defining characteristic of non-formal education is that it is an addition, alternative and/or a complement to formal education within the lifelong learning process of individuals. Another vital characteristic of non-formal education is that the education provider does not provide extra-curricular activities such as sports, to the students. Over the years, non-formal education has expanded due to dropouts and repeaters who are unable to secure places at formal institutions.

Rationale for Introducing STEM in the Zimbabwe Curriculum

With the turn of the millennium, it is noted that STEM education in the Zimbabwean curriculum was preceded by science education, which, at the primary and secondary levels, was delivered through subjects such as combined science, physics, and chemistry, culminating in advanced subjects like pure mathematics, applied mathematics, statistics, and computer science. The rationale for introducing STEM education in the teaching curriculum in Zimbabwe is to equip students with the necessary knowledge, skills, and competencies to thrive in a rapidly evolving global economy increasingly driven by technology and innovation. The key reasons for Zimbabwe incorporating STEM education are discussed in the following sections.

Economic Development

STEM fields are essential for economic growth and development. By introducing STEM education, Zimbabwe aims to produce a skilled workforce that can contribute to the country's technological advancements, innovation, and competitiveness on a global scale. Fostering STEM skills will lead to job creation, entrepreneurship, and the development of a knowledge-based economy. Zimbabwe's input into a new global economy is based on an education system biased towards STEM subjects (Ministry of Primary and Secondary Education, 2015).

Global Relevance

In an increasingly interconnected world, technological advancements and scientific discoveries significantly impact various aspects of society. By integrating STEM education, Zimbabwe ensures that its students are equipped to tackle global challenges and contribute to solutions in areas such as climate change, healthcare, energy, and sustainable development.

Future-Proofing

The jobs of the future will heavily rely on STEM skills. By introducing STEM education, the curriculum aims to provide students with a strong foundation in science, technology, engineering, and mathematics, enabling them to adapt to the changing demands of the job market. STEM education fosters critical thinking, problem-solving, creativity, and innovation, which are essential skills for future success (Bongo et al., 2016).

Enhancing Critical Thinking and Problem-Solving Skills

The competence-based curriculum framework states that schools need to engage students in STEM education that empowers them with skills such as "critical thinking and problem-solving, collaboration and leading with influence, agility and adaptability, taking the initiative and being enterprising, effective oral and written communication skills, capability to access and analyse information" (Ministry of Primary and Secondary Education, 2015, p. 11). It encourages students to ask questions, investigate, experiment, and find solutions to real-world problems. These skills are valuable in STEM-related professions and various other domains of life, helping students become lifelong learners and active contributors to society.

Promoting Innovation and Creativity

STEM education nurtures creativity and innovation by encouraging students to think outside the box, explore new ideas, and develop novel solutions. By fostering an environment that values creativity and innovation, Zimbabwe aims to cultivate a culture of entrepreneurship, scientific research, and technological advancements.

Addressing Sustainable Development Goals

STEM education can contribute to addressing the United Nations' Sustainable Development Goals (SDGs). By integrating concepts of sustainability, environmental awareness, and social responsibility into STEM education, Zimbabwe can educate students about the importance of sustainable development and empower them to make positive contributions to their communities and the world.

The goal of the STEM revolution is to establish Zimbabwe as a global leader in scientific discoveries and technological innovations, while effectively capitalizing on economic opportunities through the commercialization of research outcomes, thereby enhancing the country's competitiveness (Ministry of Higher and Tertiary Education, Science and Technology Development, 2016). Overall, introducing STEM education in the teaching curriculum in Zimbabwe is driven by the need to prepare students for the future, enhance their employability, foster innovation, and contribute to economic development. By equipping students with STEM skills, Zimbabwe aims to create a workforce that can tackle global challenges, drive technological advancements, and contribute to the country's sustainable development.

Primary to University STEM-Infused Education

In the Zimbabwean context, STEM education entails a comprehensive approach to teaching and learning these subjects from primary to university. The content covered in STEM education at each level is designed to develop students' knowledge and skills in these disciplines and prepare them for further studies or careers in STEM-related fields. Figure 1 presents the current 14-year, leveled pre-university education system, within which STEM is offered.

According to the structure of STEM outlined in Fig. 1, the Competency-Based Curriculum (CBC) 2015–2022 assumes that 12 years of education, from Grade 0 to Form 4, provides sufficient time for students to develop the necessary competencies for STEM work and literacy. Post-A-level STEM students have three options: (1) pursue STEM or non-STEM-related professional training in colleges and polytechnics, (2) seek informal or formal employment in both STEM and non-STEM fields, or (3) enroll in university STEM-related programs. The science learning path varies by level, with science and mathematics taught at the infant level, science and technology at junior primary, general science or pure sciences at O-level, and STEM/STEAM subjects at A-level. In the competence-based curriculum, STEM Education is referenced as one of the five foci of the new curriculum (Ministry of Primary and Secondary Education, 2015).

At the ECD level in Zimbabwe, STEM education focuses on introducing young children to the foundational concepts of science, technology, engineering, and mathematics in a playful and interactive manner. The goal is to foster a love for

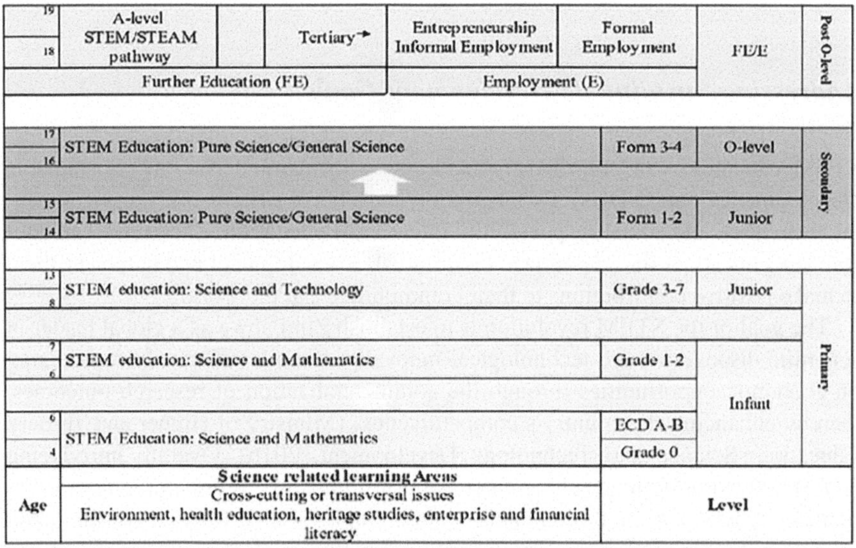

Fig. 1 Zimbabwe School Science structure (Adapted from CBC 2015–2022, p. 38)

learning, curiosity, and critical thinking skills from an early age. STEM education at the ECD level in Zimbabwe incorporates hands-on activities, games, and experiments that engage children in exploring the natural world, understanding basic mathematical concepts, and developing problem-solving abilities. The curriculum is designed to be age-appropriate and developmentally suitable for young students.

In science, children are introduced to basic scientific concepts through activities that encourage observation, exploration, and experimentation. They learn about the properties of objects, the characteristics of living things, and the concept of cause and effect. Through simple experiments and nature walks, children develop an understanding of the world around them and their scientific inquiry skills.

In Zimbabwe, ECD children are exposed to foundational mathematical concepts such as number recognition, counting, sorting, and basic shapes in mathematics. They engage in activities that involve counting objects, matching shapes, and solving simple puzzles. Children develop their numeracy skills, logical thinking, and spatial awareness through these activities. Technology is introduced at the ECD level using age-appropriate educational tools and resources. Children may be exposed to simple technology devices such as tablets or computers that offer interactive learning experiences. They may also engage in activities using basic tools, such as measuring tapes or magnifying glasses, to explore and investigate the world around them. Engineering concepts are introduced through activities that promote creativity, problem-solving, and critical thinking. Children are encouraged to build structures using blocks, create simple machines using recycled materials, and engage in imaginative play that involves problem-solving scenarios. These activities help children develop spatial skills, logical reasoning, and creativity.

In Zimbabwe's ECD level, STEM education does not emphasise formal instruction or a rigid curriculum. Instead, it focuses on a play-based and child-centred approach, where children are encouraged to explore, inquire, and discover at their own pace. Teachers and caregivers play a crucial role in facilitating these learning experiences by providing guidance and promoting a positive attitude towards STEM subjects. Overall, STEM education at the ECD level in Zimbabwe aims to lay a strong foundation for children's future learning and development in science, technology, engineering, and mathematics. By fostering a love for learning, curiosity, and critical thinking skills from an early age, STEM education at the ECD level prepares children for further studies and future success in STEM-related fields.

At the primary school level, the STEM curriculum focuses on building a strong foundation in science and mathematics (Ministry of Primary and Secondary Education, 2019). Through hands-on experiments and activities, students are introduced to basic scientific concepts, such as the properties of matter, energy, and forces. They also learn fundamental mathematical skills, including number sense, measurement, and basic operations. The aim is to develop students' curiosity, critical thinking, and problem-solving abilities from an early age.

As students progress to secondary school, the STEM curriculum becomes more specialised and in-depth. The content covered includes Mathematics, Physics, Chemistry, Biology, and Computer Science. In Mathematics, students learn advanced concepts such as algebra, geometry, and calculus. In the sciences, they delve deeper into topics like atomic structure, chemical reactions, ecosystems, and genetics. Computer Science introduces students to programming, coding, and digital literacy. Practical components, including laboratory experiments and projects, are an integral part of STEM education at this level, allowing students to apply theoretical knowledge to real-world situations.

To further enrich STEM education, Zimbabwe has established specialized STEM schools and programs at the secondary school level. These schools offer additional STEM subjects and activities, such as robotics, coding clubs, and science fairs. The curriculum in these institutions is designed to provide students with a more advanced and comprehensive understanding of STEM disciplines, preparing them for higher education and STEM careers.

At the tertiary level, universities in Zimbabwe offer a wide range of STEM-related courses and programs. Students can pursue undergraduate and postgraduate degrees in Engineering, Information Technology, Health Sciences, Applied Sciences, and Mathematical Sciences. The content covered in these programs is more specialised and advanced, focusing on theoretical knowledge, practical skills, and research. Students engage in laboratory work, research projects, and internships to gain hands-on experience and apply their knowledge in real-world settings (Ministry of Higher and Tertiary Education, Innovation, Science and Technology Development (MHTESITD), 2019). Alongside subject-specific content, STEM education in Zimbabwe also emphasises developing transferable skills like critical thinking, problem-solving, teamwork, and communication. These skills are woven into the curriculum through collaborative projects, group discussions, and presentations. The aim is to equip students with the skills and mindset required to adapt to technological advancements and contribute to innovation and economic development.

Overall, STEM education in Zimbabwe encompasses a comprehensive range of subjects and content, starting from primary school and continuing through secondary school and university. The curriculum is designed to provide students with a solid foundation in science, technology, engineering, and mathematics and develop their critical thinking, problem solving, and practical skills. Through theoretical knowledge, practical application, and hands-on experiences, STEM education in Zimbabwe aims to prepare students for further studies, research, and careers in STEM-related fields.

Outcomes and Achievements

The introduction of free and compulsory primary education since 1980 has led to a higher enrolment rate, enabling more children to enroll at school, laying a strong foundation for Science and STEM uptake. In addition, Zimbabwe is a signatory to and part of binding treaties and declarations that promote the right to education. These instruments impose an obligation on states to respect, promote, protect and fulfill the human right to education. Efforts to eliminate gender disparities have also been successful, with equal enrollment rates for boys and girls at the primary level and increasing numbers of girls pursuing secondary and tertiary education. This progress has contributed to a more inclusive STEM pipeline, with increasing female participation in traditionally male-dominated fields such as engineering, technology, and mathematics.

The education system in Zimbabwe has played a crucial role in fostering interest and participation in STEM fields. The education system has effectively equipped individuals with the fundamental literacy and numeracy skills necessary for personal and professional development. According to UNESCO, the literacy rate in Zimbabwe has steadily increased over the years, reaching approximately 89% in recent years. By providing access to education for all, regardless of socioeconomic background or geographical location, the system has created a more equitable society. Education has opened doors for disadvantaged individuals, offering them a chance to improve their circumstances, pursue higher education and break the cycle of poverty. TVET programmes have been introduced to equip students with practical skills and knowledge applicable in various STEM industries. Tertiary Institutions have also focused on producing professionals in fields such as medicine, engineering and business, contributing to economic growth and development. Through the inclusion of local languages and cultural studies in the curriculum, the education system has fostered a sense of pride and understanding of Zimbabwe's rich heritage. This has helped to build a cohesive nation that values its unique cultural identity while embracing unity and diversity. Integrating local languages and cultural perspectives into STEM education has made scientific concepts more accessible and relevant to students, fostering innovation and problem-solving skills rooted in Zimbabwe's cultural context. These efforts have created opportunities for disadvantaged students, empowering them to break socio-economic barriers and contribute meaningfully to the country's technological advancement.

The country has become a hub for educational research, training and academic collaboration, particularly in the field of STEM. Zimbabwean universities and scholars have made noteworthy contributions to research and publications, enhancing the global body of knowledge. These contributions have played a vital role in improving STEM education, fostering innovation, and addressing global challenges. Furthermore, Zimbabwean students studying abroad have gained international recognition for their academic achievements, particularly in STEM disciplines, further demonstrating the country's growing influence in these critical fields. As a result, Zimbabwe is emerging as a key player in promoting the development of STEM education, which is crucial for its socio-economic advancement.

Conclusion

The education context of Zimbabwe reflects a complex interplay of historical, social, and economic factors. Significant challenges persist while notable progress has been made in expanding access and improving educational outcomes. However, there is an ongoing commitment from the government and various stakeholders to address these challenges, provide quality STEM education and nurture its human capital. With satisfactory STEM education, Zimbabwe can secure a brighter future for its citizens and contribute to its overall social, economic, and political development.

References

Andor, L. E., Kalley, J., & Schoeman, E. (1999). *Southern African political history: A chronological of key political events from independence to mid-1997*. Bloomsbury Academic.

Atkinson, N. D. (1972). *Teaching Rhodesians: A history of educational policy in Rhodesia*. Longman.

Bongo, P. P., Bhukuvhani, C., & Gadzirayi, C. T. (2016). *Diagnostic study on the status of STEM in Zimbabwe*. Bindura University of Science Education.

Brooks World Poverty Institute. (2009). *Moving forward in Zimbabwe: Reducing poverty and promoting growth*. University of Manchester.

Chikoko, V. (2007). Negotiating roles and responsibilities in the context of decentralised school governance: A case study of one cluster of schools in Zimbabwe. *International Studies in Educational Administration, 1*(1), 22-40.

Dokora, L. D. (2015). *Minister's report on primary and secondary education to the 15th ZANU PF congress*. ZANU PF.

Education Policy and Data Center. (2014). *National education profile: Zimbabwe*. FHI160.

Hungwe, K. N. (1994). A decade of science education in Zimbabwe (1980–1990): Nationalist vision and post-colonial realities. *Journal of Curriculum Studies, 26*, 83–95.

Kanyongo, G. (2005). Zimbabwe's public education system reforms: Successes and challenges. *International Education Journal, 6*(1), 65–74.

MacKenzie, C. G. (1988). Zimbabwe's educational miracle and the problems it has created. *International Review of Education, 34*(3), 337–335.

Mapako, F. P., & Mareva, R. (2013). The concept of free primary school education in Zimbabwe: Myth or reality. *Education Research International, 1*(1), 135–147.

Matereke, K. P. (2012). 'Whipping into line': The dual crisis of education and citizenship in post-colonial Zimbabwe. *Educational Philosophy and Theory, 44*(2), 84–99.

Ministry of Higher and Tertiary Education, Science and Technology Development. (2016). *Concept note* (revised 2016, January 13). Paper Presented at the 1st national conference on Science, Technology, Engineering and Mathematics (STEM) at Harare International Conference Centre from 28–28 January. Harare: Zimbabwe.

Ministry of Higher and Tertiary Education, Science and Technology Development (2019). Strategic Plan 2019–2023. https://planipolis.iiep.unesco.org/en/2019/strategic-plan-2019-2023-7206

Ministry of Primary and Secondary Education, Zimbabwe. (2015). *Curriculum framework: Primary and secondary education 2015–2022*. Zimbabwe Ministry of Education.

Ministry of Primary and Secondary Education, Zimbabwe. (2019). ICT Policy for Primary and Secondary Education (2019–2023). MPoSE.

Moyo, J. N. (2015). *Minister's report on higher and tertiary education science and technology development to the 15th ZANU PF conference*. ZANU PF.

Mwase, T., & Marovah, T. (2023). Decolonising the Zimbabwean ordinary level food technology and design curriculum. *Cogent Education, 10*(1). https://doi.org/10.1080/2331186X.2023.2190302

Shizha, E., & Kariwo, M. T. (2011). *Education and development in Zimbabwe: A social, political, and economic analysis*. Sense Publishers.

UNICEF Zimbabwe. (2011). *Education in emergencies and post-crisis transition 2010 report evaluation*. UNICEF.

Knowledge, Curriculum, and STEM Education: Theoretical Underpinnings

Abstract This chapter offers two novel frameworks: a Framework for the Foundations of STEM Education and a Framework for Enhanced STEM Integration. The foundations of STEM education are not merely confined to curricular reforms but represent a significant paradigm shift in educational practices and philosophies. Influenced by the progressive education movement of the early twentieth century and educational philosophers like John Dewey, STEM education fosters experiential learning, problem-solving, and critical thinking. This chapter sets the stage for understanding how theoretical underpinnings can be leveraged to create educational systems that prepare students for future challenges and catalyze national development and competitiveness on the global stage. By adopting a constructivist approach, STEM education in Zimbabwe positions students as active participants in their educational journey, encouraging them to engage with and construct knowledge through meaningful interactions and experiences.

Introduction

Zimbabwe's integration into the global economy is anchored on an education system that strongly emphasises STEM (MoPSE, 2015). Similarly, the Ministry of Higher and Tertiary Education, Innovation, Science, and Technology Development (MHTEISTD) is driving the implementation of Education 5.0, which encompasses Teaching, Research, Community Service, Innovation, and Industrialisation. The aim is to propel the nation toward achieving middle-income status by 2030 (Government of Zimbabwe, 2018). STEM shapes educational practices across all levels, from basic to tertiary education. STEM disciplines undeniably drive competitive national economies (Mpofu, 2019). Globally, nations are heavily investing in STEM to nurture innovative minds capable of leading their economies' development and sustainable growth (Manokore & Sibanda, 2024). The interplay between knowledge, curriculum, and STEM education forms the foundation of modern educational systems, especially in contexts where these disciplines are considered crucial for national development and global competitiveness (Mutseekwa et al., 2024).

B. Chirinda et al., *STEM Education in Poverty*,
https://doi.org/10.1007/978-3-031-92490-3_3

The Foundations of STEM Education

STEM education, an acronym for Science, Technology, Engineering, and Mathematics, has emerged as a cornerstone in the educational landscape, reflecting the growing recognition of the pivotal role these disciplines play in driving innovation, economic growth, and global competitiveness (Mpofu, 2019; Manokore & Sibanda, 2024; Mutseekwa et al., 2024). The foundations of STEM education extend beyond curriculum adjustments, representing a paradigm shift that seeks to prepare students for the challenges of a rapidly evolving world.

Philosophical Roots of STEM Education

The philosophical underpinnings of STEM education can be traced back to the progressive education movement of the early twentieth century (Glancy & Moore, 2013). Influential educational philosophers, such as John Dewey (1966), emphasised learning by doing, hands-on experiences, and the integration of practical skills into the curriculum (Parawira, 2016). STEM education embraces these principles by prioritising experiential learning, problem-solving, and critical thinking as foundational components. The philosophical roots of STEM education emphasise an active, student-centered approach that aligns with the constructivist theory of learning (Parawira, 2016). STEM education in our Zimbabwean school system is rooted in the constructivist school of thought (Moyo, 2019). According to this philosophy, students are viewed as active participants in the learning process, capable of constructing their knowledge through their interactions and experiences. According to constructivist theories by Dewey (1966), Piaget (1947), Vygotsky (1986), and others, meaningful learning happens when students are exposed to an interactive environment where they actively engage all their senses. In a study by Madusise (2019) in Zimbabwe, some teachers acknowledge the influence of progressive education philosophers, such as John Dewey, who emphasised learning by doing and the importance of experiential learning. In Zimbabwe, STEM education is viewed as a natural extension of these principles, placing a strong emphasis on hands-on, inquiry-based activities (Madusise, 2019). Students learn best when actively engaged in the learning process, participating in experiments, projects, and real-world applications that connect theoretical concepts to practical experiences (Vygotsky, 1986).

The student-centred approach, rooted in the philosophical notion of individualised learning, resonates strongly with teachers in STEM education (Moyo, 2019). Madusise (2019) emphasised the importance of tailoring instruction to meet students' diverse needs and interests. By creating an environment where students have agency in their learning, teachers aim to foster a sense of ownership, curiosity, and intrinsic motivation as key elements in the philosophical foundations of progressive education. STEM education is aligned with the holistic development of the individual, a core tenet of the progressive education philosophy. Beyond imparting subject-specific knowledge in Zimbabwe, the curriculum emphasises the importance of nurturing critical thinking skills, creativity, and problem-solving abilities (Mabhanda, 2016). STEM subjects are viewed as vehicles for cultivating well-rounded individuals who can navigate the complexities of the modern world and contribute meaningfully to society.

The philosophical roots of STEM education highlight the enduring influence of progressive education principles on instructional practices and educational philosophy (Glancy & Moore, 2013). From experiential learning to the holistic development of an individual, teachers play a crucial role in translating these philosophical underpinnings into effective STEM education practices. STEM education can be understood through the lens of three major learning theories: information processing, constructivism, and sociocultural theory (Tenney et al., 2023). Each offers a distinct perspective on how students acquire, apply, and internalise STEM knowledge and skills.

Information Processing and STEM Education

Information Processing (IP) theory views the human mind as an information processor, similar to a computer, that manipulates abstract information (Foglia & Wilson, 2013). Knowledge is formed by encoding this information into structured rules or "if-then" statements, and learning occurs as individuals become more proficient in applying these rules (Foglia & Wilson, 2013). Teaching and learning from an IP perspective emphasize developing problem-solving proficiency, improving cognitive processing efficiency, and gradually transitioning students from novice to expert through structured instruction and practice (Tenney et al., 2023). IP theory views learning as a cognitive process in which information is received, processed, stored, and retrieved (Foglia & Wilson, 2013). In the context of STEM education, this theory emphasises the importance of structured instruction, scaffolding, and strategies that enhance students' ability to process complex STEM concepts (Tenney et al., 2023). Techniques such as chunking information, using problem-solving frameworks, and integrating digital tools can improve students' retention and application of STEM knowledge.

Constructivism and STEM Education

The learning theory of constructivism emphasises students' active role in constructing knowledge throughout the learning process (Tenney et al., 2023). Key assumptions of constructivism include the idea that individuals develop conceptual structures, or schemas, based on prior experiences (Piaget, 1947). Students acquire new knowledge by either assimilating it into existing schemas or modifying those schemas to accommodate new information (Ernest, 1995). Additionally, constructivism posits that individuals perceive the world through the lens of their own experiences rather than as an objective reality (von Glasersfeld, 1995). As students encounter new information, they interpret it in ways that align with their existing understandings, adapting their schemas as necessary. These schemas are shaped by personal experiences and influence how individuals perceive and interact with their surroundings.

The constructivist perspective on learning provides a useful framework for conceptualizing STEM education, as described by Norris and Phillips (2003), who emphasize similar principles regarding knowledge and the student. They argue that knowledge is not built linearly but is shaped by the discourse surrounding it, including verbal and nonverbal communication in the classroom, with written text playing a crucial role. This aligns with constructivist assumptions, where knowledge is formed through interconnected mental schemas rather than a rigid structure. In this view, STEM education focuses on how individuals interpret content and how texts facilitate meaning-making, reflecting the constructivist belief in a subjective reality shaped by personal experiences. Additionally, Norris and Phillips (2003) highlight that neither the teacher nor the educator holds absolute authority in the learning process, mirroring constructivism's emphasis on the student's active role in constructing knowledge rather than viewing texts as superior sources of information.

Sociocultural Learning Theory and STEM Education

The sociocultural learning theory emphasises that cognition originates in social experiences, is mediated by cultural tools, and is deeply intertwined with language and thought (Pugh, 2017). Learning occurs as individuals first engage in social practices at a collective level before internalising them for personal use (Confrey, 1995). Cultural tools, including language, are central in shaping thought and defining activities. Language, in particular, is considered the most influential mediational tool, as external speech gradually transforms into internal thought, reinforcing the connection between language and cognition (Vygotsky, 1986). From this perspective, a successful student effectively uses cultural tools to mediate thought and internalise cultural practices. In the context of STEM education, Solomon (2008) applied a sociocultural lens by framing mathematics as a social and cultural activity, asserting that mathematical literacy requires students to develop identities of participation. Participation in mathematical social practices shapes a student's understanding, and their positioning as a "math or science person" is influenced by the values, assumptions, and rules within the classroom community. These mediational tools help students interpret their roles in the learning environment. Additionally, Solomon (2008) emphasised that much of learning is linguistic, describing mathematics learning as internalising mathematical discourse to communicate effectively within and beyond the classroom. Through discourse, students develop an identity of participation, ultimately fostering STEM literacy. Having explored the

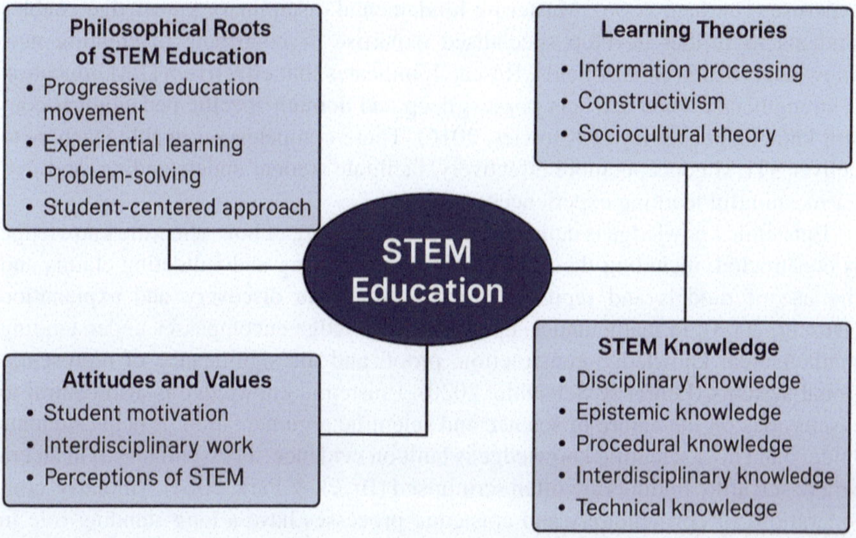

Fig. 1 Framework for the foundations of STEM education

philosophical and theoretical underpinnings of STEM education, we consolidated these ideas in a visual format. The framework in Fig. 1 summarises the key foundations: philosophical roots, learning theories, and types of STEM knowledge that inform contemporary STEM education practices.

STEM Knowledge

STEM knowledge encompasses not only disciplinary (content) knowledge but also epistemic, procedural, interdisciplinary, and technical knowledge (Ng, 2019; Tytler, 2020). In STEM education, disciplinary knowledge serves as the foundation for understanding the world, providing a structured framework through which other types of knowledge can be learned and developed. It encompasses subject-specific concepts and detailed content that students acquire within individual STEM disciplines. Disciplinary knowledge includes fundamental concepts such as energy, geometric relationships, material and structural properties, and ecosystem principles, among others (Tytler, 2020). These core concepts form the foundation of STEM education, ensuring a deep understanding of subject-specific principles that can be applied across various scientific and technological contexts. As students build disciplinary knowledge, they also gain the ability to connect ideas across different fields (interdisciplinary knowledge), understand how knowledge is applied in real-world contexts by professionals (epistemic knowledge), and engage with various processes and methods for utilising knowledge (procedural knowledge). Thus, disciplinary knowledge forms the conceptual basis for deeper understanding and expertise (Gardner, 2006). Mastering fundamental disciplinary knowledge enables students to further develop specialised expertise or contribute to creating new knowledge within STEM fields. Research indicates that effective STEM education is strengthened when teachers possess deep and domain-specific pedagogical content knowledge (Kelley & Knowles, 2016). These competencies enable teachers to deliver STEM concepts more effectively, facilitate student understanding, and foster meaningful learning experiences.

Epistemic knowledge is defined as an understanding of how scientific knowledge is constructed, including the role of evidence in testing and validating claims and the use of models and representations in scientific discovery and explanation (Duschl, 2008). In mathematics, epistemic knowledge encompasses understanding mathematical knowledge construction, proof, and the significance of representational systems (Lehrer & Schauble, 2020). Epistemic knowledge is also central to discussions on the nature of science and scientific argumentation. It helps students understand how scientific knowledge is built on evidence, an essential skill in an era where scientific findings are often scrutinised (Brock & Park, 2024). Similarly, considerations of epistemology and epistemic processes have a long-standing role in mathematics education (Ernest, 2003), reinforcing the importance of understanding how mathematical and scientific knowledge evolves. Epistemic knowledge also encompasses understanding the personal and social influences that drive STEM discovery and development and the diverse professional and practical contexts in which STEM practitioners operate (Tytler, 2020). This aligns with calls for STEM education to more accurately reflect real-world STEM practices more accurately, provide students with exposure to authentic STEM experiences, and foster collaborations between schools, STEM industries, and professionals (Tytler, 2020).

Procedural knowledge forms the foundation for acquiring and applying STEM skills, such as measuring data, assessing its precision, validity, and reliability, and selecting and presenting it effectively (Ng, 2019). It is developed through investigative and hands-on activities inside and outside the classroom and should be introduced progressively to align with evolving scientific and technological advancements (Tytler, 2020). Students can independently explore STEM principles and their applications by fostering procedural knowledge.

Interdisciplinary knowledge has not traditionally been a central focus in mathematics and science education. However, mathematics serves as a critical tool in scientific inquiry, while applied mathematics often relies on scientific contexts (Tytler, 2020). Similarly, technology and engineering design projects integrate mathematics and science, although typically in a utilitarian manner, without deep conceptual engagement (Joseph & Uzondu, 2024). There is a growing recognition that interdisciplinary thinking is a fundamental aspect of contemporary STEM professions, with innovation frequently emerging at the intersection of disciplinary boundaries. This highlights the need to explore structured approaches for fostering interdisciplinary learning in mathematics and science education (Parawira, 2016; Tytler, 2020). Industry stakeholders advocate for this integration, recognising that creativity, design, and interdisciplinary collaboration are essential to STEM practices (Tytler, 2020; Joseph & Uzondu, 2024).

Technical knowledge in STEM education is fundamentally linked to applying knowledge, skills, attitudes, and values within specific fields, careers, or tasks (Ng, 2019). This includes disciplines such as civil engineering, where theoretical understanding is applied to create practical solutions to real-world challenges. According to Ng (2019), technical knowledge bridges the gap between conceptual learning and professional practice, equipping students with the necessary competencies to innovate and solve complex problems within STEM-related industries. Industry and government increasingly view STEM education as a key platform for developing complex problem-solving, critical thinking, and analytical and quantitative reasoning skills, all highly valued in the modern workforce (Tytler, 2020; Nkala et al., 2024). As a result, the STEM curriculum must prioritise cognitive and metacognitive skill development to ensure students can navigate and contribute to evolving professional and societal demands.

A crucial outcome of STEM education is the ability to apply knowledge to interpret and solve real-world problems (Svodziwa, 2021). This shift toward "knowledge in use" aligns with broader educational goals such as preparing all citizens for informed participation in society and equipping students with the adaptable skills and competencies needed in the future workplace (Tytler, 2020).

Attitudes and values are increasingly recognised as essential learning components and are critical in sustaining student engagement with STEM subjects (Tytler et al., 2023). In particular, attitudes toward mathematics and science are key drivers of interdisciplinary work in schools, shaping students' motivation and long-term participation in STEM education. Values influence students' perceptions of STEM and are explicitly promoted within STEM curricula (Schreiner & Sjøberg, 2007; Bishop et al., 2003). A key aspect of mathematical learning is a productive disposition, defined as the tendency to view mathematics as a sensible, useful, and worthwhile subject, combined with a belief in the value of hard work and self-efficacy in problem-solving (Kilpatrick et al., 2001). This competency is crucial for learning mathematics and science and applying STEM knowledge in everyday life and professional contexts (Tytler, 2020).

Approaches to Curriculum STEM Integration

Literature has highlighted various models and approaches reflecting the complexity of STEM education conceptualisation (Mutseekwa et al., 2024). Many proponents of STEM education argue that the best way to teach STEM subjects is through an integrative approach, which may include interdisciplinary teaching and learning that bridges STEM disciplines with non-STEM subject areas (Moore & Smith, 2014). Mutseekwa et al.'s (2024) analysis identified multiple frameworks for STEM education integration, all addressing the challenge of merging STEM disciplines. Kertil and Gurel (2016) proposed two primary models: content integration, which structures curricula to incorporate multiple STEM disciplines, and context integration, which centres on one discipline while drawing relevant contexts from others without compromising depth and rigour. Johnson et al. (2015) identified three types of classroom STEM integration: content integration, supporting content integration, and context integration. Content integration involves activities that meet multiple STEM learning objectives. Supporting content integration addresses one STEM content area's objectives to support another area's learning goals. Context integration uses the context of one STEM discipline to support teaching in another. STEM lesson activities designed for integration should include real-world problem solving, engineering design, appropriate technological applications supported by standards-based math and science, and content from other disciplines like English/ language arts and social studies (Moore et al., 2016). The integrative approach fosters increased interest in mathematics, improves students' attitudes toward learning, and demonstrates the real-world applications of mathematics. Research has shown that integrative teaching approaches enhance mathematics learning and strengthen STEM knowledge, preparing students for college and STEM-related careers (Kertil & Gürel, 2016).

Strimel and Grubbs (2016) outlined three additional models: (1) maintaining technology education's fundamental goals, (2) merging engineering and science education to form a core discipline, and (3) revitalising science through an engineering-oriented approach. These models emphasise different aspects of STEM education, with engineering often serving as a mediating discipline due to its design-focused methodology. However, a critical challenge remains: teachers' understanding of these integration approaches, as effective STEM implementation requires specialized pedagogical content knowledge (Thibaut et al., 2018). Integrative STEM design incorporates multidisciplinary, interdisciplinary, and transdisciplinary teaching and learning approaches in English (2017). STEM integration can occur at four varying levels (disciplinary, multidisciplinary, interdisciplinary and transdisciplinary), each reflecting a different approach to the relationship between disciplines (Vasquez et al., 2013). Figure 2 shows the levels of STEM integration.

A Disciplinary or Discipline-Specific Approach

A disciplinary or discipline-specific approach in STEM curriculum refers to the traditional model where each STEM subject (Science, Technology, Engineering, and Mathematics) is taught as a separate and distinct entity (Mpofu, 2019). In this approach, the focus is on subject-specific content and methodologies within each discipline. Students learn the foundational concepts, principles, and skills relevant to each area without necessarily making explicit connections across the disciplines. The approach often prioritises depth of knowledge in each subject, with limited emphasis on interdisciplinary connections or integrating the knowledge and practices of different STEM fields (Mpofu, 2019). While it allows students to develop expertise within individual disciplines, the approach may not fully facilitate the development of cross-disciplinary skills or the ability to apply knowledge in real-world, integrated contexts.

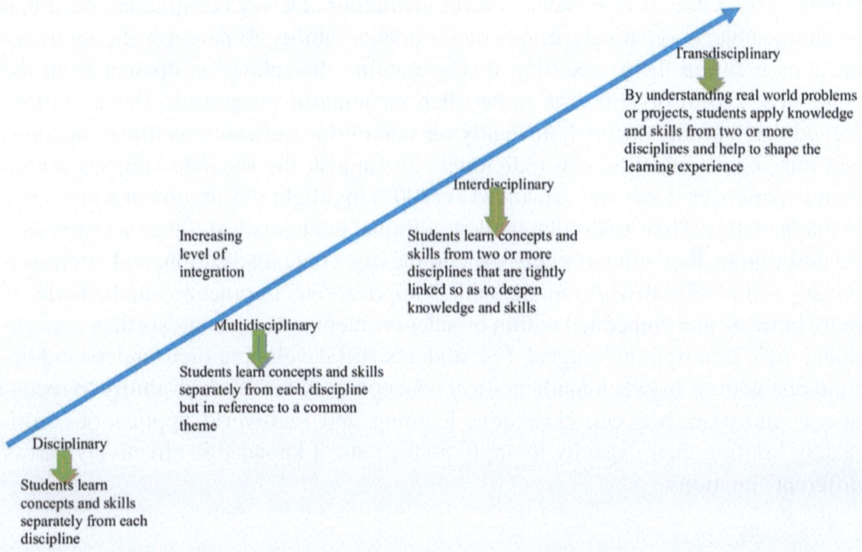

Fig. 2 A continuum of STEM approaches to curriculum integration (Vasquez et al., 2013)

A discipline-specific approach is predominantly adopted in secondary schools, whereas greater integration is often observed at the primary level (Ng, 2019). Typically, all students engage in science and mathematics. At the same time, technology and engineering-related subjects such as Design and Technology, ICT, Computer Science, and Electrical Engineering are often offered as electives, particularly in secondary education. In some cases, STEM-related vocational courses are also available at the secondary level.

Under this discipline-specific approach, teachers continue instructing their respective subjects, with STEM functioning as an umbrella term rather than a fully integrated framework (Ng, 2019). Each discipline remains distinct and separate, even though each inherently incorporates scientific, mathematical, engineering, technological, and design thinking (Mpofu, 2019). While these disciplines can independently contribute to holistic STEM education, their interconnections may not be explicitly emphasised, potentially limiting the acquisition of cross-disciplinary competencies (Ng, 2019). To encourage student participation in STEM fields at the tertiary level or STEM-related careers, subject combinations are sometimes strategically packaged to guide students in selecting courses that best align with future STEM learning and professional pathways (IBE-UNESCO, 2013).

Moore et al. (2016) emphasise that students need to be prepared to address societal challenges by providing them with rich, engaging, and interdisciplinary STEM experiences. They argue that separating STEM disciplines creates artificial divides that do not accurately reflect real-world contexts, whereas integration presents these fields more authentically and practically. This perspective aligns with Dewey's

(1966) early critiques of isolated subject instruction. Dewey (1966) contended that teaching subjects separately diminishes students' ability to perceive the interconnections between them, asserting that presenting disciplines as distinct from the outset leads to disorganisation rather than meaningful integration. Dewey (1966) argued that learning occurs holistically outside of the artificial structure of schools, and only upon reflection can individuals distinguish the separate subjects within their experiences. Lesh and Zawojewski (2007) highlight this argument's relevance to mathematics. They assert that while traditional mathematical topics serve as useful descriptors, they often obscure the complexity, contextualisation, and interdisciplinary nature of real-world mathematical applications. In practice, mathematics is more intricate and embedded within broader problem-solving contexts than conventional topic descriptions suggest. For students still developing their understanding, rigid distinctions between mathematical concepts may hinder their ability to recognise connections between classroom learning and real-world applications, ultimately limiting their capacity to apply mathematical knowledge effectively across different situations.

Interdisciplinary or Integrated STEM Education Curriculum

Interdisciplinary or integrated STEM education is a fusion of the four disciplines (Science, Technology, Engineering, and Mathematics) where the primary focus shifts from individual subject areas to solving real-world problems (Ng, 2019). The changing demands of the 4IR, which continuously reshape the global labour market, underscore the need for multifaceted, trans-disciplinary, and integrated competencies (Marope, 2017). The integrated STEM education curriculum helps guide students to think creatively and critically, enabling them to identify problems and construct innovative solutions. Research indicates that an interdisciplinary curriculum offers more relevant, less fragmented, and engaging learning experiences (Ng, 2019). It improves student motivation, enhances attitudes towards learning, and equips students with vital skills such as problem-solving, innovation, self-reliance, logical thinking, and technological literacy (Stohlmann et al., 2012). Integrated STEM education, which connects the subjects to real-world challenges, further strengthens these competencies.

Multidisciplinary STEM integration represents the lowest level of integration, while transdisciplinary STEM integration is the highest (Wang & Knobloch, 2018). These levels differ in three key aspects. First, at the multidisciplinary level, Science, Technology, Engineering, and Mathematics (STEM) identities remain distinct, with content and skills taught separately. In contrast, transdisciplinary integration blurs or eliminates disciplinary boundaries by engaging students in real-world problems or projects. Second, the lowest level of integration is more teacher-centred, where educators lead instruction and determine the content and skills to be taught (Mpofu, 2019). At the highest level, STEM integration becomes more student-centred, with teachers acting as facilitators while students take an active role in deciding the content, skills, and direction of their learning. Finally, the focus of learning shifts across these levels: multidisciplinary integration emphasises discipline-specific concepts and skills, whereas transdisciplinary integration fosters the application of knowledge and skills across disciplines within real-world contexts (Mpofu, 2019).

STEME Integrated Framework

Scholars have established that integrated STEM education is recognised as a field (Johnson et al., 2020; Jackson et al., 2021; Roehrig et al., 2021). Mpofu (2019) developed a Science, Technology, Engineering and Mathematics Education (STEME) (see Fig. 3) framework, arguing that education is contextual as STEM.

According to Mpofu (2019), the lowest level of STEM integration is the separatist approach, abbreviated as S–T–E–M, where the hyphens signify that each STEM subject is taught separately rather than in an integrated manner. In this approach, integration is achieved simply by adding a STEM subject to the existing school curriculum. Literature indicates that this method is one way to incorporate technology into school curricula (Mpofu, 2019). Under a basic conception of STEM education encompassing technology, engineering, and mathematics, schools typically add the

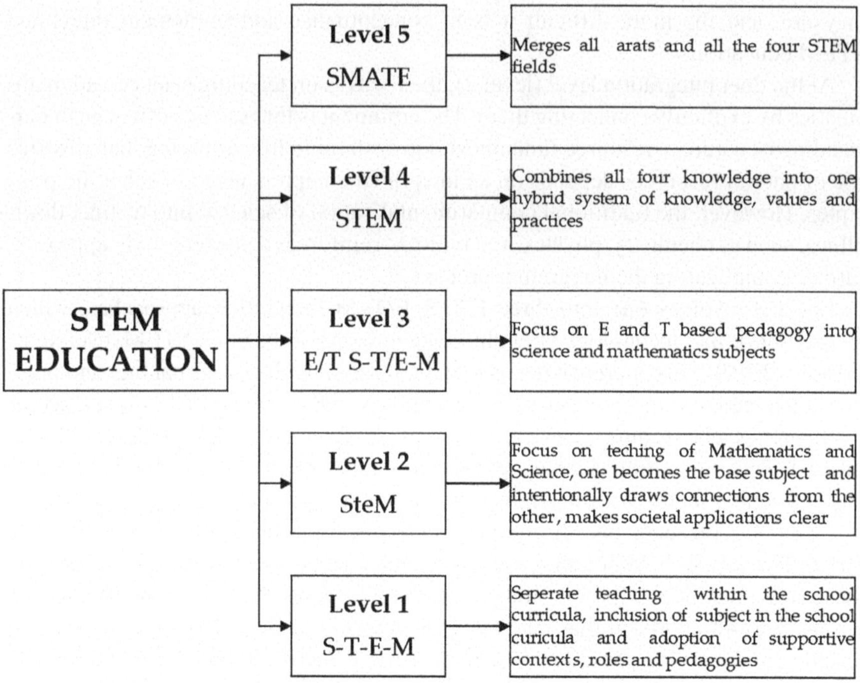

Fig. 3 STEME framework (Mpofu, 2019)

missing discipline, often technology and engineering, to the curriculum. This straightforward model of STEM education is relatively easy to implement, as it primarily requires the training of engineering and technology teachers.

The traditional separatist approach upholds a long-standing philosophy of subject specificity and disciplinary separation. Today, its influence can be seen in policies that aim to increase the number of STEM subjects, courses, or academic programs in educational institutions. These policies often target issues such as declining enrollment and gender disparities within STEM fields. However, while some advocate for increasing the number of STEM offerings, others argue that proper pedagogical integration should occur within the teaching of individual STEM subjects. For instance, many countries, including Zimbabwe, promote a shift in pedagogical practices. The new STEM curriculum framework in Zimbabwe encourages teachers to adopt research-based, discovery-oriented, and problem-based approaches when teaching specific STEM subjects. Such student-centred strategies, particularly in STEM subjects, are expected to foster creative and innovative thinking among students.

The STEME level 2–4 approaches refer to forms of STEM integration involving more than one subject (Mpofu, 2019). Generally, these integration methods are more challenging to understand and implement than the separatist approach. These approaches are ordered based on the logic that establishing connections among STEM subjects becomes increasingly complex as more subjects are involved. In other words, the greater the number of subjects, the more cognitively demanding they are, and the more difficult it is to conceptualise and implement integrated STEM education.

At the duet integration level (level 2), the focus is on teaching science and mathematics by explicitly connecting them. The argument is that science offers an understanding of nature, a resource fundamental to sustaining life, implying that effective use of natural resources depends on an integrated comprehension of scientific principles. However, the traditional compartmentalisation of science into distinct disciplines, such as chemistry, physics, and biology, reinforces a discrete study approach, further complicating the integration process.

Level 3 involves one into three E/T S–E/T–M integration approaches, which incorporate either technology or engineering into one of the other STEM disciplines (Mpofu, 2019). This approach posits that the reasoning, decision-making, and practices characteristic of engineering or technology can be integrated into science or mathematics classrooms (or vice versa), depending on which subject is being infused. This integration is pedagogically grounded, as it involves adopting design-based teaching methods from engineering or technology within science instruction.

Level 4 represents interdisciplinary or multidisciplinary integration (Mpofu, 2019). Even though Although some studies have attempted to distinguish between these two constructs, Mpofu's study treats them as equivalent, referring to interdisciplinary as an approach that integrates all four STEM disciplines into a cohesive teaching and learning paradigm. The main goal of interdisciplinary integration is to break down the traditional paradigmatic barriers that separate the four STEM disciplines. Within this framework, teachers are expected to help students make connections between academic concepts in science, technology, engineering, and mathematics and real-world contexts, thereby linking school, community, work, and the global enterprise. Moreover, this approach to interdisciplinary learning has been shown to positively influence lifelong learning habits, enhance academic skills, promote personal growth, and develop effective knowledge management skills. Research indicates that integrated STEM instruction is more effective than teaching disciplines separately (Jackson et al., 2021; Tytler et al., 2023). Integrating disciplines bridges the theoretical knowledge of science and mathematics with the practical applications found in engineering and technology (Mpofu, 2019). Moreover, Jackson et al. (2021) contend that inquiry-based project learning is a cornerstone of STEM education that not only develops quantitative and scientific literacy but also cultivates the twenty-first-century competencies essential for success in the STEM workforce.

The final (level 5) approach is called Mathematics, Science, Art, Technology, and Engineering (MSATE), a modification of the traditional STEAM model. While STEAM education generally signifies the addition of the Arts to STEM (i.e., STEM + Arts), a model initially developed for students with disabilities, the MSATE framework is adapted for all students. In this approach, the arrangement of letters represents the integration of diverse knowledge bodies into a holistic system of knowledge, values, skills, and practices. This configuration echoes Indigenous scholars' arguments that science is inherently cultural and that language is central to every culture's knowledge development. The art domain, which embraces languages and the social sciences, is placed at the centre to signify its crucial role in knowledge development, representation, and communication. Mathematics and science precede art, highlighting mathematics' role as a connective thread across STEAM disciplines and science's fundamental role in understanding nature. In this way, the linguistic understanding inherent in mathematics and science supports their combined application in technology and engineering, acknowledging that language use is integral to educational activities within social contexts (Mpofu, 2019).

Towards a Framework for Enhanced STEM Integration

The continuum of STEM approaches to curriculum integration as presented in Fig. 2, provides a conceptual and pedagogical classification of STEM integration. It identifies various teaching and curriculum models ranging from disciplinary, to multidisciplinary, interdisciplinary, and transdisciplinary. It categorizes integration by content and context, emphasizing classroom practice and pedagogical depth. However, this framework lacks a sociocultural and contextual lens, omitting the ways in which integration relates to local education systems, teacher capacities, resource availability, and student needs.

The second framework in Fig. 3 (Mpofu's STEME framework, 2019) emphasizes levels of integration, placing STEM education within the African schooling contexts like Zimbabwe. It contextualizes integration levels in relation to practical realities and incorporates the MSATE approach, which aligns with inclusive and cultural awareness in education. However, Mpofu's model is hierarchical and linear, focusing more on curriculum structure and less on instructional strategies, real-world applications, and feedback loops between content and pedagogy. In response to these limitations, we propose a Framework for Enhanced STEM Integration.

The Framework for Enhanced STEM Integration (Fig. 4) synthesizes the pedagogical classifications of STEM integration with the sociocultural and contextual insights offered by Mpofu's (2019) STEME model. It bridges the theoretical models and classroom realities by incorporating levels of integration, content-context alignment, and culturally responsive pedagogy. The framework also highlights feedback loops between teaching practice and curriculum design, encouraging dynamic and inclusive STEM education strategies that are contextually grounded and pedagogically sound.

The Framework for Enhanced STEM Integration reflects a deliberate progression from low to high levels of integration, starting from disciplinary (separatist) approaches and moving through duet, one-into-three, interdisciplinary/multidisciplinary, and culminating in the transdisciplinary level embodied by the MSATE model. This layering effectively illustrates how the complexity and cognitive demand increase as the integration deepens.

The inclusion of real-world application, engineering design, contextual learning, and student-centeredness at the higher levels aligns well with global trends in education reform and the demands of the 4IR. Notably, the central placement of STEM education within this enhancement continuum underscores its role as a transformative pillar rather than a static set of subjects. The visual flow suggests that proper STEM education is not merely about content inclusion but also about pedagogical innovation, epistemological alignment, and cultural contextualisation.

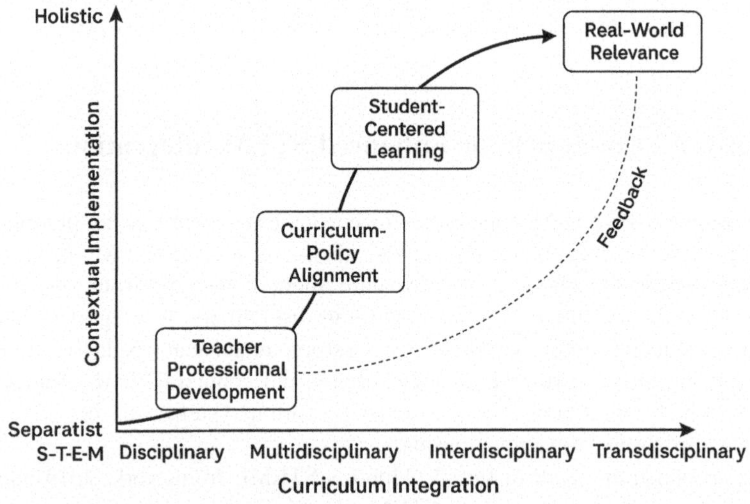

Fig. 4 Framework for enhanced STEM integration

One notable enhancement to this framework is the inclusion of a feedback loop, which represents a critical feature of dynamic, responsive curriculum development in STEM education. The feedback loop signifies the iterative process of curriculum refinement, teacher professional development, and student learning outcomes. It ensures that insights from implementation, such as student engagement, assessment results, and contextual challenges, are systematically fed back into the planning and design phases. This cyclical process enables continuous improvement of STEM pedagogical strategies and ensures alignment with evolving educational goals, technological advancements, and societal needs. Incorporating a feedback loop transforms the framework from a linear or static representation into a more fluid, systems-thinking model that reflects the dynamic nature of teaching and learning. The feedback mechanism emphasizes the importance of ongoing reflection, evaluation, and adjustment based on practical classroom experiences, student outcomes, and evolving contextual demands. It supports continuous professional development for teachers, adaptation of curriculum content, and responsiveness to student needs, thereby fostering a sustainable and inclusive approach to STEM integration. The loop also underscores the value of stakeholder input, ensuring that policy, practice, and pedagogy remain interconnected and mutually reinforcing throughout the implementation process.

Moreover, the framework bridges the gap between theory and practice by connecting academic models with actionable curriculum design considerations. It helps stakeholders visualize how to progress from fragmented teaching to integrated and culturally inclusive instruction that prepares students for real-world challenges. Overall, the framework functions as both a conceptual map and a strategic tool for educators, curriculum developers, and policymakers striving to implement meaningful and effective STEM education.

Conclusion

Despite these theoretical models, implementing integrated STEM education within traditionally compartmentalised school structures presents significant challenges (Mutseekwa et al., 2024). Transitioning from a segregated, discipline-based system to an integrated STEM approach requires comprehensive curriculum restructuring and a fundamental shift in instructional design (Thibaut et al., 2018). English (2017) emphasised that advancing STEM integration requires balancing core disciplinary content and interdisciplinary approaches. Strong STEM curricula should promote twenty-first-century skills such as inquiry, problem-solving, critical thinking, creativity, and innovation while upholding a robust foundation in disciplinary knowledge (Roehrig et al., 2021). STEM education is not merely a collection of disciplines but a pedagogical approach that integrates these fields to solve real-world problems. Its success depends on the effective design and implementation of curricula aligning with knowledge's epistemological foundations.

References

Bishop, A., Seah, W. T., & Chin, C. (2003). Values in mathematics teaching: The hidden persuaders? In *Second international handbook of mathematics education* (pp. 717–765). Springer.

Brock, R., & Park, W. (2024). Distinguishing nature of science beliefs, knowledge and understandings. *Science and Education, 33*, 495–516. https://doi.org/10.1007/s11191-022-00368-6

Confrey, J. (1995). A theory of intellectual development. *For Learning Mathematics, 15*, 36–45.

Dewey, J. (1966). In R. Arachambault (Ed.), *Lectures in the philosophy of education: 1899*. Random House.

Duschl, R. (2008). Science education in three-part harmony: Balancing conceptual, epistemic, and social learning goals. *Review of Research in Education, 32*(1), 268–291.

English, L. D. (2017). Advancing elementary and middle school STEM education. *International Journal of Science and Mathematics Education, 15*, 5–24. https://doi.org/10.1007/s10763-017-9802-x

Ernest, P. (1995). The one and the many. In L. P. Steffe & J. Gale (Eds.), *Constructivism in education* (pp. 459–486). Erlbaum.

Ernest, P. (2003). *Constructing mathematical knowledge: Epistemology and mathematics education*. Routledge.

Foglia, L., & Wilson, R. A. (2013). Embodied cognition. Wiley interdisciplinary reviews. *Cognitive Science, 4*(3), 319–325.

Gardner, H. (2006). *Five minds for the future*. Harvard Business School Press.

Glancy, A. W., & Moore, T. J. (2013). *Theoretical foundations for effective STEM learning environments*. School of Engineering Education Working Papers. Paper 1. Retrieved from http://docs.lib.purdue.edu/enewp/1

Government of Zimbabwe. (2018). *Doctrine education 5.0 heritage, innovation, industrialisation. The modernisation & industrialisation of Zimbabwe through science education & development. Harare: Ministry of Higher and Tertiary Education, Science and Technological Development [MoHTEISTD].*

IBE-UNESCO. (2013). *IBE glossary of curriculum terminology*. IBE-UNESCO. Retrieved from https://unesdoc.unesco.org/ark:/48223/pf0000223059

Jackson, C., Mohr-Schroeder, M. J., Bush, S. B., Maiorca, C., Roberts, T., Yost, C., & Fowler, A. (2021). Equity-oriented conceptual framework for K-12 STEM literacy. *International Journal of STEM Education, 8*(1), 38. https://doi.org/10.1186/s40594-021-00294-z

Johnson, C. C., Peters-Burton, E. E., & Moore, T. J. (Eds.). (2015). *STEM road map: A framework for integrated STEM education* (1st ed.). Routledge. https://doi.org/10.4324/9781315753157

Johnson, C., Mohr-Schroeder, M. J., Moore, T., & English, L. (2020). *Handbook of research on STEM education*. Routledge.

Joseph, O. B., & Uzondu, N. C. (2024). Curriculums development for interdisciplinary STEM education: A review of models and approaches. *International Journal of Applied Research in Social Sciences, 6*(8), 1575–1592.

Kelley, T. R., & Knowles, J. G. (2016). A conceptual framework for integrated STEM education. *International Journal of STEM Education, 3*, 11. https://doi.org/10.1186/s40594-016-0046-z

Kertil, M., & Gürel, C. (2016). Mathematical modelling: A bridge to STEM education. *International Journal of Education in Mathematics, Science and Technology, 4*(1), 44–55. https://doi.org/10.18404/ijemst.95761

Kilpatrick, J., Swafford, J., & Findell, B. (2001). *Adding it up: Helping children learn mathematics*. National Academy Press.

Lehrer, R., & Schauble, L. (2020). Stepping carefully: Thinking through the potential pitfalls of integrated STEM. *Journal for STEM Education Research, 4*, 1–26. https://doi.org/10.1007/s41979-020-00042-y

Lesh, R. A., & Zawojewski, J. (2007). Problem solving and modeling. In F. K. Lester (Ed.), *Second handbook of research on mathematics teaching and learning* (pp. 763–804). Information Age Publishing.

Mabhanda, W. (2016). Opportunities and factors affecting adoption of STEM education: The case of Gweru Polytechnic first year commerce students. *International Journal of Business Marketing and Management, 1*(5), 1–8.

Madusise, S. (2019). Possibilities for implementation of STEM education in Zimbabwean underresourced mathematics classrooms. *Journal of African Interdisciplinary Studies, 3*(7), 110–124.

Manokore, K., & Sibanda, L. (2024). National STEM education framework: Teachers' perspectives on the 2015–2022 curriculum cycle. *Journal of Research in Education and Pedagogy, 1*(2), 98–106. https://doi.org/10.70232/jrep.v1i2.12

Marope, M. (2017). Future competences for future generations. In S. Popa (Ed.), *IBE in focus* (pp. 81–89). IBE-UNESCO.

Moore, T., & Smith, K. (2014). Advancing the state of the art of STEM integration. *Journal of STEM Education. Innovations and Research, 15*(1), 5–10.

Moore, T., Johnson, C., Peters-Burton, E., & Guzey, S. (2016). The need for a STEM road map. In *STEM road map* (pp. 4–10). Routledge.

MoPSE. (2015). *Curriculum framework: Primary and secondary education 2015–2022*. Zimbabwe Ministry of Education.

Moyo, M. (2019). Towards implementing a STEMitised curriculum in inclusive secondary school contexts: The case of five selected schools in Mzilikazi District of Bulawayo. *Zimbabwe Journal of Educational Research, 31*(3), 470-484.

Mpofu, V. (2019). *A theoretical framework for implementing STEM education. Theorizing STEM education in the 21st century*. Intech Open.

Mutseekwa, C., Dzavo, J., Musaniwa, O., & Nshizirungu, G. (2024). Framing pre-service teacher preparation in Africa from global STEM education practices. *Pedagogical Research, 9*(3), em0215. https://doi.org/10.29333/pr/14701

Ng, S. (2019). *Exploring STEM competences for the 21st century. In-progress reflection No. 30 on current and critical issues in curriculum, learning and assessment*. United Nations Educational, Scientific and Cultural Organisation (UNESCO).

Nkala, B., Sibanda, V. M., Ndhlovu, J., Hobane, L., & Singh, S. (2024). Challenges and drivers of industrialist propensity among chemical engineering students in STEM institution in Zimbabwe: Towards a conceptual framework. *International Journal of Environment, Engineering and Education, 6*(1), 26–37. https://doi.org/10.55151/ijeedu.v6i1.126

Norris, S. P., & Phillips, L. M. (2003). How literacy in its fundamental sense is central to scientific literacy. *Science Education, 87*, 224–240. https://doi.org/10.1002/sce.10066

Parawira, W. (2016). Science in our natural world. *The Sunday Mail*. Retrieved from www.Sundaymail.co.zw

Piaget, J. (1947). *The psychology of intelligence*. Routledge.

Pugh, K. J. (2017). *Computers, cockroaches, and ecosystems: Understanding learning through metaphor*. Information Age Publishing.

Roehrig, G. H., Dare, E. A., Ring-Whalen, E., & Wieselmann, J. R. (2021). Understanding coherence and integration in integrated STEM curriculum. *International Journal of STEM Education, 8*(2). https://doi.org/10.1186/s40594-020-00259-8

Schreiner, C., & Sjøberg, S. (2007). Science education and youth's identity construction—Two incompatible projects? In D. Corrigan, J. Dillon, & R. Gunstone (Eds.), *The re-emergence of values in the science curriculum* (pp. 231–247). Sense Publishers.

Solomon, Y. (2008). *Mathematical literacy: Developing identities of inclusion*. Routledge.

Stohlmann, M., Moore, T. J., & Roehrig, G. H. (2012). Considerations for teaching integrated STEM education. *Journal of Pre-College Engineering Education Research (J-PEER), 2*(1), Article 4. https://doi.org/10.5703/1288284314653

Strimel, G., & Grubbs, M. E. (2016). Positioning technology and engineering education as a key force in STEM education. *Journal of Technology Education, 27*(2), 21–36. https://doi.org/10.21061/jte.v27i2.a.2

Svodziwa, M. (2021). *STEM education as a human capital development strategy for the industrialization and modernization of Zimbabwe*. Retrieved from https://ssrn.com/abstract=3788852 or https://doi.org/10.2139/ssrn.3788852

Tenney, K., Stringer, B. P., LaTona-Tequida, T., & White, I. (2023). Conceptualizations and limitations of STEM literacy across learning theories. *Journal Microbiology Education, 24*(1), 1–5. https://doi.org/10.1128/jmbe.00168-22

Thibaut, L., Ceuppens, S., De Loof, H., De Meester, J., Goovaerts, L., Struyf, A., Boeve-de Pauw, J., Dehaene, W., Deprez, J., De Cock, M., Hellinckx, L., Knipprath, H., Langie, G., Struyven, K., Van de Velde, D., Van Petegem, P., & Depaepe, F. (2018). Integrated STEM education: A systematic review of instructional practices in secondary education. *European Journal of STEM Education, 3*(1), 2. https://doi.org/10.20897/ejsteme/85525

Tytler, R. (2020). STEM education for the twenty-first century. In J. Anderson & Y. Li (Eds.), *Integrated approaches to STEM education. Advances in STEM education*. Springer. https://doi.org/10.1007/978-3-030-52229-2_3

Tytler, R., Anderson, J., & Williams, G. (2023). Exploring a framework for integrated STEM: Challenges and benefits for promoting engagement in learning mathematics. *ZDM Mathematics Education, 55*, 1299–1313. https://doi.org/10.1007/s11858-023-01519-x

Vasquez, J., Sneider, C., & Comer, M. (2013). *STEM lesson essentials, grades 3–8: Integrating science, technology, engineering, and mathematics*. Heinemann.

von Glasersfeld, E. (1995). *Radical constructivism: A way of knowing and learning*. Falmer Press.

Vygotsky, L. S. (1986). *Thought and language, revised education*. MIT Press.

Wang, H.-H., & Knobloch, N. A. (2018). Levels of STEM integration through agriculture, food, and natural resources. *Journal of Agricultural Education, 59*(3), 258–277. https://doi.org/10.5032/jae.2018.03258

STEM Education Pedagogy in Developing Country Contexts: Relevance and Challenges

Abstract This chapter examines the significance and implementation of STEM education pedagogy in developing countries, using Zimbabwe as a case study. The need for STEM education in developing nations is driven by economic diversification and the transition to knowledge-based economies, necessitating proficiency in science and technology. The chapter outlines the importance of the four Cs of twenty-first-century education—Creativity, Collaboration, Critical Thinking, and Communication—and emphasizes the necessity of an inquiry-based, adaptable framework to foster a love of learning among students. Despite the high demand for STEM professionals, developing countries face challenges such as resource allocation, outdated infrastructure, and cultural barriers that impede the growth of a skilled workforce. A survey of 798 Zimbabwean educators was conducted to provide valuable insights into the current practices of educators in implementing STEM pedagogy across Zimbabwe. The survey revealed a reliance on traditional teaching methods in Zimbabwe's STEM education system, with notable disparities between rural and urban settings in implementing more innovative techniques.

The Vital Role of Effective STEM Pedagogy in Developing Countries

STEM embraces the four Cs identified as key in twenty-first-century education: Creativity, Collaboration, Critical Thinking, and Communication. Most importantly, by incorporating inquiry-based principles and a highly adaptable framework to suit students' various needs, STEM helps foster a love of learning. Estimates indicate that over 30.6 million unskilled Africans cannot take advantage of STEM opportunities in the continent. Except for India, no developing countries rank among the top 15 nations with the highest number of STEM degrees. Consequently, job offers for qualified workers vastly outnumber the available employees by hundreds of thousands. This imbalance causes businesses to suffer, progress in many sectors to slow down, and developing countries to bear the highest costs. Nonetheless,

B. Chirinda et al., *STEM Education in Poverty*,
https://doi.org/10.1007/978-3-031-92490-3_4

effective STEM pedagogy plays a crucial role in fostering innovation and economic growth. By prioritizing effective pedagogy in STEM education in developing countries, we can equip students with the essential skills and knowledge, they need to tackle real-world challenges. This approach not only enhances individual learning but also supports broader community development and technological advancement. Investing in effective STEM pedagogy paves the way for a brighter future, benefiting both individuals and society as a whole.

A glimmer of optimism arises from the awareness that STEM professions are growing in popularity and importance in many developing countries, thanks to technology's continued vital role in driving economic growth and development. However, as STEM professions move from being "popular" in the minds of young people to being "popular" among university students, optimism must face reality. The reasons for the shortage of technical and scientific professionals vary from region to region. STEM jobs require advanced degrees that require more time and financial resources to complete than training in other professions. In developing countries, the reasons vary from living conditions and mobility in conflict zones and those affected by natural disasters to the demographic boom, which directs resources to different priorities, such as student infrastructure, such as campuses, classrooms, laboratories, outdated equipment, and technology. Cultural reasons also penalise women's emancipation, not to mention hierarchical dynamics and family expectations that guide young people's educational and career choices away from their passions and interests.

Research indicates that the economic performance of metropolises with more STEM-oriented economies is higher than that of those with less investment in STEM subjects. Unemployment is lower, incomes are higher, and patents for innovative products and services benefit from suitable public-private-academic synergies that allow them to flourish. Many experts also believe this is true at the national level: the most prosperous countries in the world tend to use scientific advances and the latest technologies efficiently. Experts also argue that a STEM education encourages critical thinking, promotes voluntary in-depth exploration of academic subjects, fosters collaboration, and stimulates creativity and problem-solving skills. It is indeed breakthroughs in science and technology, driven by a workforce skilled in science, technology, engineering, and mathematics (STEM), that will enable Africa to overcome crippling developmental challenges, including climate change, food insecurity, inequality, and poverty.

Teaching STEM subjects helps students develop essential life skills such as critical thinking, problem solving, innovation, teamwork, and confidence. One challenge teachers face is the ongoing need for creative approaches to maintain students' interest in STEM. European educators and industrialists have identified a widening gap in STEM skills within the workforce (Kelley & Knowles, 2016). Furthermore, the enhancement of STEM education is driven by economic issues in developed and emerging countries (Kennedy & Odell, 2014). The Fourth Industrial Revolution marks a new era of economic growth characterized by an understanding of STEM fields such as big data, artificial intelligence, the Internet of Things, robotics, and 3D printing. The demand for professionals who integrate knowledge of mathematics, physical sciences, digital technologies, chemistry, biology, and other disciplines is rising. Students studying various fields such as industrial biotechnology, food science, artificial intelligence, data science, engineering, and robotics must understand diverse technical and physical processes, employ critical thinking and problem-solving approaches, analyse data and process flows, and integrate different interdisciplinary perspectives.

STEM Pedagogy in Developing Countries

Pedagogy in developing countries refers to the teaching methods, strategies, and approaches used in educational systems to facilitate learning and promote student development (Westbrook et al., 2013). This chapter explores STEM pedagogy in developing countries, with a particular focus on Zimbabwe, highlighting the instructional methods used. Understanding these dynamics is essential for developing strategies to improve STEM education and ensure that students acquire the necessary skills to contribute to scientific and technological advancements in their respective countries.

STEM Teaching and Learning Approaches in Zimbabwe

Various approaches are recommended for STEM education at different levels in Zimbabwe. The diverse approaches aim to strengthen STEM education and equip students with the skills needed for the evolving job market and technological advancements. Figure 1 illustrates the STEM pedagogy framework, which outlines the core teaching and learning approaches adopted across various educational levels within Zimbabwe's STEM education landscape.

Early Childhood Development

ECD STEM education in Zimbabwe emphasizes a play-based approach, where young children engage in hands-on activities and exploration to develop foundational STEM skills. Play is a widely used teaching pedagogy in ECD classrooms, recognized for its positive effects on cognitive and social development. Through play, young children unconsciously gather and process information, develop new skills, and practice them by engaging in activities organized by adults (Pakombwele, 2023). Research indicates that play is essential for developing various skills during early childhood (Pakombwele, 2023), noting that play fosters discovery learning, a fundamental aspect of early childhood education. Play supports holistic development, as Pakombwele (2023) suggests that children's natural curiosity drives their

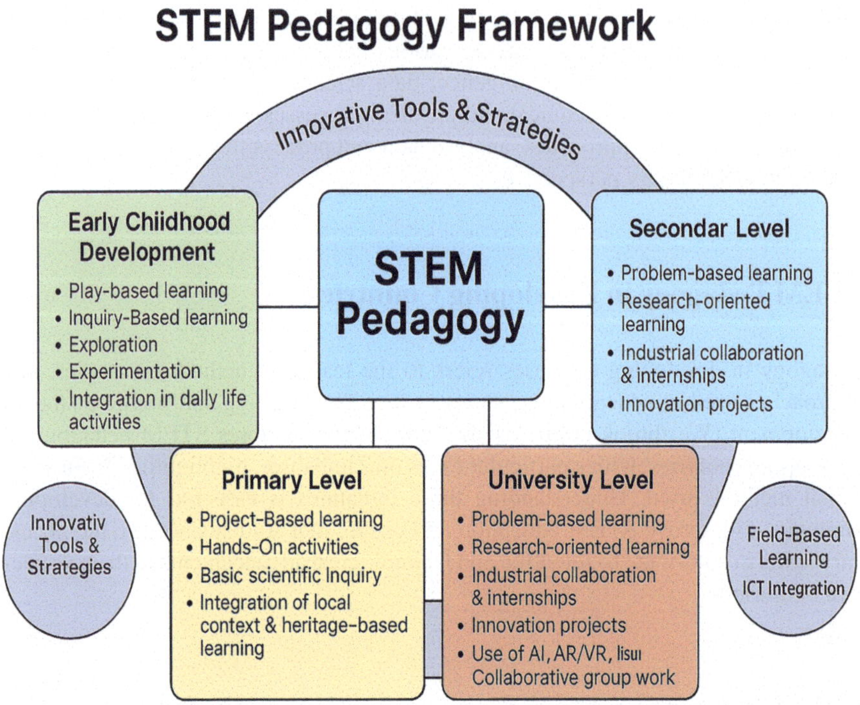

Fig. 1 STEM pedagogy framework

exploration of their surroundings. Consequently, play is the foundation for teaching and learning in ECD classrooms, enhancing memory and recall skills, particularly in mathematics and science. Furthermore, play provides young children with opportunities to interact, explore, practice skills, develop problem-solving abilities, and regulate emotions, all of which are crucial for cultivating scientific skills in ECD students (Pakombwele & Tsakeni, 2024).

Encouraging curiosity and critical thinking through inquiry-based learning approaches to help children develop problem-solving and scientific inquiry skills is emphasised at the ECD level in Zimbabwe. Inquiry-based pedagogy is a learning approach where the quest for knowledge, opinions, observations, and critical thinking are central to the educational process (Mugweni, 2016). This method posits that both teachers and students share responsibility for learning, with teachers actively engaging in the process and fostering a participatory culture that encourages students to express and refine their ideas collaboratively. In ECD, this approach involves open-ended explorations that promote student autonomy during practical experiments (Tsakeni et al., 2019). The teacher acts as a guide, helping students explore their thoughts and develop their inquiries to logical conclusions. This method is particularly relevant in teaching mathematics and science in ECD classrooms, as it places students at the forefront of their educational experiences (Pakombwele, 2023; Pakombwele & Tsakeni, 2024). In the context of the twenty-first century, where problem-solving, critical thinking, and investigative skills are vital, inquiry-based learning allows students to engage deeply with their environment rather than merely memorizing isolated facts (Pakombwele, 2023). As Mugweni (2016) asserts, the most effective way to learn science is through active investigation, data collection, and questioning, as students construct knowledge by interacting with their surroundings.

Experimentation is stressed in ECD classrooms in Zimbabwe. Young children are naturally curious and eager to experiment, making early childhood the ideal time to foster a love for science (Pakombwele, 2023; Pakombwele & Tsakeni, 2024).) suggest that by offering children ample hands-on experiences, their curiosity is further stimulated. These experiences help them develop essential skills such as collaboration, communication, perseverance, teamwork, problem-solving, and analytical reasoning. Classroom experiments expose students to scientific concepts that encourage critical thinking.

In contrast, the basic scientific and mathematical vocabulary they acquire helps them make sense of the world and understand how things work. According to Pakombwele (2023), experiments are vital for young children as they allow them to seek answers to their questions and develop broader concepts. In ECD classrooms, experiments play a key role in developing science skills.

Exploration is encouraged in ECD classrooms in Zimbabwe. Exploration is a method that ECD teachers use when teaching science, as contemporary research highlights its importance in early childhood development (Pakombwele & Tsakeni, 2024). According to Pakombwele (2023), exploring science is an exciting part of early childhood, and children are natural-born scientists, making early exploration key to developing lifelong problem-solving skills. As children are naturally curious, they investigate and discover how the world works, and through exploration activities, they develop important skills such as communication, problem-solving, and investigation. During experimentation, children interact with materials and peers to construct their knowledge, and when they engage in activities such as questioning, experimenting, and observing, they maximize their brain's learning capacity (Pakombwele, 2023).

Integrating STEM into daily activities is important at the ECD level in Zimbabwe. STEM concepts can be integrated into everyday activities, such as counting and sorting objects, exploring nature, and building with blocks, to foster early STEM learning. The Ministry of Primary and Secondary Education (MoPSE, 2015) highlighted in its curriculum document that the revised approach, rooted in heritage-based education, emphasizes the practical application of learning to real-life situations within cultural contexts. Specifically, education should be guided by the concept of "ubuntu/unhu/vumunhu," which represents universal human interdependence, solidarity, compassion, and a sense of community—values central to African societies (MoPSE, 2015, p. 13).

Primary Level

In Zimbabwe, primary-level STEM education emphasises project-based learning, which involves practical, hands-on activities that engage students actively in the learning process. Implementing project-based learning approaches allows students to tackle on real-world problems, thereby fostering creativity, collaboration, and problem-solving skills. Chavunduka (2017) further emphasizes the use of project-based methods in teaching science, encouraging hands-on, practical learning in schools. The competence based STEM curriculum introduced coursework projects, which were very welcome to promote and enhance authentic learning.

Secondary Level

Zimbabwe promotes the integration of STEM subjects into the secondary school curriculum, providing students with a strong foundation in science, technology, engineering, and mathematics. The competence-based curriculum prioritises the teaching and learning of STEM subjects in schools (MoPSE, 2015). Practical work, including laboratory experiments and hands-on activities, is emphasised to enhance students' understanding of STEM concepts and develop practical skills.

The competence-based STEM curriculum introduced some practical tasks, which are very welcome to promote and enhance authentic learning. Student-centred practical approaches to teaching and learning are preferred in the Zimbabwean science curricula, as exemplified by the competence-based curriculum syllabi for Biology, Chemistry, and Physics (Ministry of Primary and Secondary Education, 2015a, 2015b, 2015c). The competence-based A-Level science syllabi (2015–2022) emphasize that teaching and learning should be student-centred, incorporating practical approaches such as experimentation, demonstrations, simulations, and projects. These methods encourage students to plan, design, and conduct investigations (Ministry of Primary and Secondary Education, 2015a, 2015b, 2015c, 2015d, 2015e). The O-Level Biology syllabus (Zimbabwe School Examinations Council, 2011a, 2011b, p. 2) referenced "a practical and investigative approach… in teaching this syllabus," which can be interpreted as advocating for a hands-on, minds-on approach aligned with constructivist principles. Prior to 2011, all science syllabi emphasized the importance of providing a valuable educational experience through well-designed experimental and practical science studies. For instance, the Chemistry syllabi (University of Cambridge Local Examinations Syndicate, 1993, pp. 3, 26) highlighted practical work as a key method for active knowledge construction. The aims outlined in syllabi before and after 2015 suggest that classroom practical experiences should cultivate both practical and generic skills that are transferable and applicable beyond the scope of science disciplines and careers.

Encouraging the formation of STEM clubs and participation in STEM competitions provides students with additional opportunities for exploration, innovation, and teamwork. In Zimbabwe, STEM clubs and competitions have become an important platform for promoting STEM education among students. For instance, the National High Schools Quiz competition, which includes STEM-related questions, encourages critical thinking and scientific inquiry. Additionally, the Zimbabwe Science Fair, held annually, provides students with the opportunity to showcase innovative projects that address real-world problems using STEM principles. In many schools, STEM clubs focus on hands-on activities, such as robotics, coding, and renewable energy projects. For example, STEM clubs that engage students in building solar-powered devices, encouraging sustainable innovation. These initiatives help nurture creativity and problem-solving skills, aligning with the country's drive to industrialize through the development of homegrown scientific knowledge.

The competence-based curriculum framework in Zimbabwe emphasizes the need for teachers to shift their pedagogical approaches. Teachers are now seen as "co-explorers and facilitators in knowledge discovery, guiding students to an objective understanding of content while demonstrating acquired skills" (Ministry of Primary and Secondary Education, 2015, p. 41). As a result, teaching and learning are student-centered, characterized by meaningful learning, holistic self-development (including self-confidence and interpersonal skills), and fostering self-awareness of learning potential and challenges, with teachers serving as facilitators, moderators, and mentors. The framework also highlights classroom interaction, active participation, and a positive, supportive, and safe learning environment.

The curriculum framework highlights several specific pedagogical practices, such as the discovery method, project-based learning, problem-based learning, and design-based learning. These approaches carry important implications for STEM education in Zimbabwe, where teaching methods should focus on creating environments that enable students to explore and solve real-life mathematical problems collaboratively. The recommended pedagogical strategies across various STEM syllabi include guided discovery, cooperative learning/group work, interactive e-learning, problem-solving, discussion, modeling, simulation, exposition, demonstration, educational tours, field trips, case studies/research and project-based learning (Ministry of Primary and Secondary Education, 2015b, 2015c). These approaches align with broader educational goals of STEM education, ensuring consistency in how mathematics is taught and how learning is structured.

A closer analysis of these pedagogical practices reveals that the teaching and learning of STEM subjects should center around problem-solving and investigations of real-life situations. STEM problems rooted in real-world contexts must be developed and utilized as tools for teaching specific STEM concepts and topics.

University Level

Universities in Zimbabwe often adopt problem-based learning approaches in STEM education, where students work on complex, real-world problems and develop analytical and critical thinking skills. Students individually or collaboratively identify a problem, analyze it, explore possible solutions, and ultimately create a scaled-down or full-size artefact (Nezandonyi, 2023). For instance, they might be tasked with developing an environmentally friendly source of light energy for use in the home. These projects help students develop skills in research, time management, human behavior, as well as managing finances and resources (Nezandonyi, 2023).

Emphasizing research and independent inquiry allows university students to deepen their understanding of STEM disciplines and contribute to new knowledge creation. At the university level in Zimbabwe, STEM research-oriented learning plays a pivotal role in fostering innovation and addressing local challenges. Universities such as the University of Zimbabwe and the National University of Science and Technology (NUST) have integrated research into their STEM curricula, allowing students to apply theoretical knowledge to real-world problems. For example, at NUST, engineering students have been involved in projects focused on renewable energy, including the design of solar-powered water pumping systems for rural communities. These research-oriented programs not only enhance students' analytical and problem-solving skills but also contribute to Zimbabwe's broader goal of promoting industrialization and innovation through STEM education.

Collaborations with industries and offering internships provide Zimbabwean university students with practical experience and exposure to real-world STEM applications. Harare Institute of Technology (HIT) students collaborate on design projects in teams, conducting research and developing prototype solutions to real-world problems. These projects provide opportunities for students to work alongside peers with diverse abilities and beliefs, thereby fostering both teamwork and communication skills. One notable project, HIT 300, involves students individually tackling design and innovation challenges during a 30-week internship or industrial training placement in established companies. This experience enables them to apply theoretical principles and ethics learned on campus to real-life situations, while planning innovative solutions. In addition to technical skills, students also develop essential soft skills such as teamwork, problem-solving, and self-discipline, which are crucial for preparing for industry settings (Nezandonyi, 2023).

Innovative Tools and Strategies in STEM Pedagogy in Zimbabwe

STEM pedagogy constantly seeks innovative tools and strategies to enhance students' learning experiences, which is evident in developing countries. As technology advances rapidly, educators find new ways to engage and empower students in STEM subjects. The STEM subjects syllabus for 2015–2022 states that the main teaching methods are Experiments, Demonstrations, Problem-Solving, Field trips, Games, Cooperative learning/Group work, Simulations, and Case studies/research (Zimbabwe Ministry of Primary and Secondary Education, 2015). These methods are best implemented using ICT tools.

One innovative tool that has gained prominence in the past decade (2000–2010) is virtual reality (VR) technology. VR provides a unique and immersive learning experience, allowing students to explore and interact with complex STEM concepts in a virtual environment. For example, students residing in developing countries can dissect virtual organisms, simulate experiments in physics or chemistry, or even explore the vastness of the universe. Using VR, students in developing countries can visualise abstract concepts, manipulate variables, and gain a deeper understanding of the underlying principles.

Another innovative tool is augmented reality (AR), which overlays digital information onto the real world. AR allows students to blend digital content with their physical surroundings, bringing abstract STEM concepts to life. For instance, students can use AR to visualise complex mathematical formulas or explore the inner workings of a human body by scanning a textbook page. AR facilitates a more interactive and immersive learning experience by engaging multiple senses, making seemingly complex topics more accessible and engaging.

Educators in developing countries can motivate students and make learning more enjoyable by incorporating game elements like competition, rewards, and levels. Gamified learning platforms allow students to solve STEM-related puzzles, engage in simulations, or participate in virtual challenges. This not only fosters a sense of excitement but also encourages critical thinking, problem-solving, and collaboration skills. Students should be given opportunities to engage in live or virtual games, individually or in groups, followed by presenting oral or written reports on the games played. For example, games like Living Loop Systems, which illustrate how overstocking can lead to soil erosion, can be used to foster an understanding of this concept. Such games stimulate students' imagination, encouraging creativity as they actively participate or observe. Dewey emphasised that meaningful learning happens through play or enacting ideas, thereby reinforcing the educational value of such activities.

Educators can incorporate Artificial Intelligence (AI) into their pedagogy, which is known to promote personalised learning in STEM. AI systems analyse student data, identify knowledge gaps, and provide tailored learning pathways. Adaptive platforms can generate customised quizzes and feedback based on individual student needs. By adapting the learning outcomes to each student's level and pace,

AI-driven systems optimise learning outcomes and ensure that students receive targeted support.

Educators can utilize collaborative tools and strategies that facilitate effective teamwork while enhancing communication and problem-solving skills. Students can collaborate on coding projects, design engineering solutions, or conduct scientific research. This collaborative approach boosts creativity and teamwork, exposing students to diverse perspectives and real-world challenges, thus preparing them for future STEM careers.

A Survey on the Reality of STEM Implementation in Zimbabwe

While various interactive and student-centered teaching methods are recommended, their implementation in Zimbabwe's schools and higher education institutions often varies. A survey was conducted to evaluate the extent to which lecture-based instruction, project-based learning, problem-solving methods, practical demonstrations, computer simulations, field-based instruction, technology integration, and language of instruction are implemented in STEM education across primary and secondary schools (rural, urban, and private), as well as in universities and colleges in Zimbabwe. Participants rated the frequency of these approaches on a five-point Likert scale (Never, Rarely, Sometimes, Often, Very Often) and provided qualitative feedback on the challenges they face in implementing these methods. A quantitative research design was used, employing structured questionnaires as the primary data collection tool. The survey targeted 798 participants, consisting of 700 primary and secondary school teachers (400 from rural schools, 200 from urban schools, and 100 from private schools) and 98 STEM teacher educators (42 from universities and 56 from teacher training colleges).

Findings

The survey provided valuable insights into the current practices by educators in implementing STEM pedagogy across Zimbabwe.

Predominance of Lecture-Based Instruction

The survey found that lecture-based instruction remains the dominant teaching method across all educational levels in Zimbabwe. Seventy-five percent of primary and secondary school teachers reported that they mainly rely on lectures in their

STEM lessons, with 80% of university and college educators also frequently using lecture-based methods. The use of lectures was particularly common in rural schools (85%), where teachers cited large class sizes, a lack of laboratory facilities, and inadequate access to digital resources as significant barriers to adopting more interactive methods. Despite these challenges, lecture-based instruction offers a structured approach, although it often fails to engage students actively in hands-on STEM learning.

Project-Based Learning Implementation

Project-based learning is widely recommended to foster problem-solving and critical thinking, but its implementation remains inconsistent. Only 15% of rural school teachers reported regularly using project-based learning, compared to 40% in urban schools and 65% in private institutions. Rural teachers identified overcrowded classrooms, lack of time, and resource constraints as significant obstacles. At the university and college level, 50% of STEM educators reported integrating project-based learning, particularly in engineering, applied sciences, and teacher training courses. However, rigid curricula and insufficient institutional support hindered the full integration of this approach across disciplines.

Adoption of Problem-Solving Methods

Problem-solving is an essential component of STEM education, but its implementation varies significantly across different settings. Fifty-five percent of urban and private school teachers actively incorporated problem-solving tasks into their lessons, compared to 30% of rural school teachers. Rural educators cited the exam-centric curriculum and time constraints as significant barriers to using problem-solving methods. At the university level, 80% of STEM educators used problem-solving methods, particularly in mathematics, engineering, and research-based courses. However, some lecturers observed that students entering higher education lacked adequate problem-solving skills, indicating gaps in secondary education preparation.

Use of Practical Demonstrations

Practical demonstrations are integral to STEM education, but their use largely depends on the availability of resources. Sixty percent of private school teachers and 50% of university educators reported regularly using practical demonstrations in their lessons. In contrast, only 25% of rural school teachers frequently conducted hands-on experiments due to limited laboratory facilities and a lack of materials. Many rural teachers expressed frustration about relying on theoretical explanations instead of hands-on experiments. In universities, practical demonstrations were frequently used in physics, chemistry, and biology courses, but outdated equipment and limited space hindered their effectiveness.

Integration of Computer Simulations

Using computer simulations in STEM education is one of the least implemented strategies in Zimbabwe. While 70% of universities and colleges have incorporated computer simulations in engineering, physics, and medical sciences, only 30% of urban and 10% of rural schools have access to these digital tools. Rural teachers cited a lack of computers, poor internet connectivity, and insufficient training as key obstacles. In contrast, 55% of private schools have integrated computer simulations, primarily due to the availability of well-equipped ICT labs and smaller class sizes. At the university level, educators emphasize the value of simulations in visualizing complex STEM concepts but acknowledge that many students entering higher education lack exposure to these tools.

Educational Field Trips and Field-Based Instruction

Field-based learning through educational field trips was more common in urban and private schools than in rural ones. Fifty percent of private and urban school teachers reported using field trips as part of their STEM instruction, while only 20% of rural teachers used field-based learning. Rural teachers cited financial constraints and transportation challenges as key barriers. At the university level, 40% of STEM educators incorporated field-based instruction, especially in applied sciences, engineering, and agricultural studies, although logistical and financial constraints limited the frequency of such experiences.

Integration of Technology in STEM Education

The survey found that technology integration was highly inconsistent across the Zimbabwean education system. Seventy percent of private and urban school teachers reported using various technologies, such as interactive whiteboards, educational software, and online resources, to enhance their STEM teaching. However, only 25% of rural teachers used technology in their classrooms, citing unavailability of electricity, lack of computers, poor internet connectivity, and limited training as significant obstacles. At the university level, 80% of STEM educators integrated technology into their teaching, particularly in courses related to engineering, computer science, and medical fields. However, technology integration was often more advanced in universities, where adequate funding, infrastructure, and access to digital tools were available.

English as the Dominant Language of Instruction

The survey data indicated that English is the dominant language of instruction in STEM subjects in Zimbabwe, with 90% of secondary and university educators reporting its use in STEM classrooms. Primary school teachers also acknowledged the transition to English as the medium of instruction after the first few years of schooling, with 75% reporting that students are expected to grasp English to engage effectively with the STEM curriculum.

The prevalence of English in STEM education affects teaching methods. Many educators, particularly those in rural schools, have expressed the need to simplify technical language and provide additional support to students who face language barriers. Sixty percent of rural school teachers indicated that they often use local languages to explain complex STEM concepts, especially during the initial learning stages. Urban teachers encounter fewer challenges as their students generally possess a stronger foundation in English and have greater access to supportive resources, including English-language textbooks and online materials.

Language barriers intersect with technology access in STEM education. As digital tools and computer simulations increasingly become integral to teaching, students who lack proficiency in English may struggle to navigate English-language software or educational platforms. Fifty percent of rural school teachers reported that students often face difficulties using technology-driven resources, not only due to limited access but also because of the language of instruction. This challenge limits the effectiveness of technology in enhancing STEM learning, particularly for rural students with lower English proficiency. At the university level, the use of English in STEM courses is less of a challenge, but it still creates difficulties for students from marginalized or non-English-speaking backgrounds. Seventy-five percent of university educators indicated that language proficiency impacts students' academic performance, especially in scientific writing and technical explanations.

Discussion

The findings revealed significant disparities in the implementation of STEM pedagogy across Zimbabwe. While private schools and universities have made strides in integrating interactive, technology-driven teaching methods, rural schools face considerable challenges due to limited resources and insufficient teacher training.

STEM provides a hands-on, immersive learning experience that requires advanced technology and high-tech gadgets for effective implementation (Mufanechiya & Makgalwa, 2024). This situation forces developing countries, including Zimbabwe, to struggle with providing adequate resources and infrastructure for STEM education, such as a shortage of well-equipped laboratories, computers, and internet access, which obstructs hands-on learning experiences and access to information (Kadziya & Ndebele, 2020). STEM implementation relies on the availability of electricity, a central component in the current industrial revolution, which is often lacking in many local schools, particularly in remote areas (Muyambo-Goto et al., 2023). Therefore, the absence of electricity prevents the recommended use of technologies such as computers and the internet that facilitate STEM education in the Fourth Industrial Revolution.

The survey highlights the dominance of lecture-based instruction, which is compounded by infrastructure constraints, as the main method of teaching STEM subjects, particularly in rural areas. Traditional rote learning approaches in Zimbabwe often prioritise academic achievements over practical application. A study by Ruparanganda (2019) in Zimbabwean schools and Mutseekwa (2021) in teachers' colleges showed that the traditional lecture method is widely used in the teaching and learning of STEM subjects. This usually results in students lacking an understanding of how STEM concepts relate to real-world problems thereby limiting their ability to apply their knowledge meaningfully.

Hands-on experiences and practical applications are essential in STEM education. However, in Zimbabwe, the limited availability of internships, industry collaborations, and practical experiences restricts students' exposure to real-world STEM scenarios and hinders their ability to connect theoretical knowledge to practical application (Mutseekwa, 2021). While methods such as project-based learning, problem-solving, and field-based instruction are recommended, their implementation is often constrained by exam-focused curricula and financial limitations. Zimbabwe can create a more inclusive and effective STEM education system that better equips students to meet the challenges of the twenty-first century.

Despite Zimbabwe having 16 official languages and many others in developing countries, the language of instruction is not typically the students' first language (Chimbunde & Kgari-Masondo, 2023). In Zimbabwe, indigenous languages are used as the medium of instruction only for the first 5 years of education, after which English becomes the primary language of instruction. To qualify for tertiary education, students must pass English at the Ordinary Level, a requirement inherited from the colonial education system (Sibanda & Young, 2019). This creates additional challenges for students' understanding of complex STEM concepts, resulting in limited comprehension and lower engagement with the subject matter.

STEM pedagogy in Zimbabwe, as outlined in the intended curriculum, promotes interactive and hands-on learning approaches such as project-based learning, problem-solving methods, computer simulations, practical demonstrations, and field-based instruction. The curriculum emphasizes the integration of real-world applications, interdisciplinary learning, and technology-driven teaching to enhance students' understanding and engagement in STEM subjects. However, the implemented curriculum often relies heavily on lecture-based instruction with limited practical engagement. While universities and some schools incorporate experiments and digital tools, the extent of their use varies, particularly between well-resourced urban and private schools and under-resourced rural schools. Additionally, emphasising an examination-oriented curriculum often reduces inquiry-based and exploratory learning opportunities, leading to a gap between the intended and implemented STEM education approaches. Despite these challenges, efforts continue to enhance STEM pedagogy by integrating more interactive and experiential learning strategies to ensure students develop the critical thinking and problem-solving skills necessary for scientific and technological advancement.

Concluding Remarks

In Zimbabwe, STEM Education attempts to transform the typical teacher-centred classroom by encouraging a curriculum driven by problem-solving, discovery, and exploratory learning. It requires students to actively engage in a situation to find its solution (Gadzirayi et al., 2016). It is important to note that these approaches may vary across different schools and institutions in Zimbabwe. Implementing these approaches also depends on resources, infrastructure, and teacher training. Continuous efforts are being made to improve STEM education at all levels in Zimbabwe, focusing on providing high-quality learning experiences and preparing students for future STEM careers.

Our document analysis revealed that some documents in Zimbabwe guide changes in pedagogy for teaching science in Zimbabwean schools. For example, the O-Level Combined Science syllabus and the textbook "Step Ahead Combined Science Learner's Book" by Chavunduka (2017) advocate for a science pedagogy rooted in community contexts. The syllabus highlights the importance of selecting examples and teaching contexts that align with students' interests, enabling them to expand and reframe their knowledge in ways that embrace diversity. In addressing the transformation of Zimbabwe's education curricula, the second Science, Technology, and Innovation (STI) policy (2012) advocates for the development of courses on Indigenous Knowledge Systems (IKS) for inclusion in the school curriculum. It also promotes research into the potential applications of IKS to address future national development challenges, complementing existing and emerging technologies. This policy framework suggests that Indigenous Knowledge should be integrated across all learning areas and subjects. Incorporating students' prior knowledge into STEM-oriented curricula fosters innovative thinking and constructivism, as students draw from their socio-cultural context and understand their local environment.

Furthermore, our document analysis revealed that the Zimbabwean curriculum emphasises using the environment as an essential component of STEM education, promoting hands-on learning experiences that enhance students' understanding of scientific concepts. The curriculum encourages teachers to integrate the local environment into instruction to provide real-world applications of STEM principles

(Madusise, 2019). This approach allows students to explore scientific and mathematical concepts through interaction with natural and built environments, thereby fostering inquiry-based learning and developing problem-solving skills. Teachers are encouraged to use field-based instruction, educational trips, and outdoor experiments to illustrate STEM concepts, making learning more engaging and relevant.

References

Chavunduka, K. (2017). *Step ahead combined science. Form 4 learner's book.* Pearson Education Africa.

Chimbunde, P., & Kgari-Masondo, M. C. (2023). The decolonisation of African languages: Insights from Southern Africa. *Language Matters, 53*(3), 5–24. https://doi.org/10.1080/10228195.2022.2138957

Gadzirayi, C. T., Bongo, P. P., Ruyimbe, O., Bhukuvhani, C., & Mucheri, T. (2016). *Diagnostic study on the status of STEM education in Zimbabwe*. Bindura University of Science Education and Higherlife Foundation.

Kadziya, L., & Ndebele, C. (2020). Towards improving the quality of teaching and learning of STEM subjects in rural post-colonial Zimbabwe. *iAfrican Renaissance, 17/i*(4), 141–165.

Kelley, T. R., & Knowles, J. G. (2016). A conceptual framework for integrated STEM education. *International Journal of STEM Education, 3*, 11. https://doi.org/10.1186/s40594-016-0046-z

Kennedy, T. J., & Odell, M. R. L. (2014). Engaging students in STEM education. *Science Education International, 25*(3), 246–258.

Madusise, S. (2019). Possibilities for implementation of STEM education in Zimbabwean under-resourced mathematics classrooms. *Journal of African Interdisciplinary Studies, 3*(7), 110–124.

Ministry of Primary and Secondary Education. (2015a). *Biology syllabus. Form 5–6: 2015–2022.* Curriculum Development and Technical Services. MoPSE.

Ministry of Primary and Secondary Education. (2015b). *Chemistry syllabus. Form 5–6: 2015–2022.* Curriculum Development and Technical Services. MoPSE.

Ministry of Primary and Secondary Education. (2015c). *O-Level chemistry syllabus. Form 3–4: 2015–2022.* Curriculum Development and Technical Services. MoPSE.

Ministry of Primary and Secondary Education. (2015d). *O-Level Physics syllabus. Form 3–4: 2015–2022.* Curriculum Development and Technical Services. MoPSE.

Ministry of Primary and Secondary Education. (2015e). *Physics syllabus. Form 5–6: 2015–2022.* Curriculum Development and Technical Services. MoPSE.

MoPSE. (2015). *Handbook on teacher professional standards*. MoPSE.

Mufanechiya, A., & Makgalwa, M. M. (2024). Re-visioning Zimbabwe's Science, Technology, Engineering and Mathematics (STEM) for secondary schools through the science teacher education programme. *Journal of Education and Learning Technology, 5*(6), 111–129. https://doi.org/10.38159/jelt.2024564

Mugweni, R. M. (2016). Early childhood (ECD) teachers' conceptions and implementation of the child-centred approach to teaching science. *International Journal of Academic Research in Progressive Education and Development, 5*(4), 2226–6348.

Mutseekwa, C. (2021). STEM practices in science teacher education curriculum: Perspectives from two secondary School teachers' colleges in Zimbabwe. *Journal of Research in Science, Mathematics and Technology Education, 4*(2), 75–92. https://doi.org/10.31756/jrsmte.422

Muyambo-Goto, O., Naidoo, D., & Kennedy, K. (2023). Students' conceptions of 21st century education in Zimbabwe. *Interchange, 54*, 49–80. https://doi.org/10.1007/s10780-022-09483-3

Nezandonyi, J. (2023). Embedding education for sustainable development in the teaching of technology courses in Zimbabwe: Policy and practice. In B. Chirinda, L. Sibanda, J. Vere, & G. Sunzuma (Eds.), *Science, Mathematics and Technology education in Zimbabwe: Research, policy and practice* (pp. 159–189). Peter Lang.

Pakombwele, A. (2023). Early childhood development in science and mathematics education in Zimbabwe: Challenges and issues. In B. Chirinda, L. Sibanda, J. Vere, & G. Sunzuma (Eds.), *Science, Mathematics and Technology education in Zimbabwe: Research, policy and practice* (pp. 159–189). Peter Lang.

Pakombwele, A., & Tsakeni, M. (2024). Designing learning centres that facilitate knowledge construction on science process skills in early childhood classrooms. *Journal of Ecohumanism, 3*(7), 755–764. https://doi.org/10.62754/joe.v3i7.4241

Ruparanganda, F. (2019). Challenges faced by Biology Secondary teachers when using instructional methods that promote Science Technology, Engineering and Mathematic (STEM) Education in Zimbabwe. *Texila International Journal of Academic Research, 6*(2), 1–6.

Sibanda, L., & Young, J. (2019). Towards a postcolonial curriculum in Zimbabwe: A critical re view of the literature. *Africa Education Review, 17*, 50–72. https://doi.org/10.1080/18146627.2018.1507622

STI. (2012). *Second Science, Technology, and Innovation policy of Zimbabwe*. Ministry of Science and Technology Development.

Tsakeni, M., Vandeyar, S., & Potgieter, M. (2019). Inquiry opportunities presented by practical work in school physical sciences. A South African case study. *Gender and Behaviour, 17*(3), 13722–13733.

University of Cambridge Local Examinations Syndicate. (1993). *Chemistry*. University of Cambridge.

Westbrook, J., Durrani, N., Brown, R., Orr, D., Pryor, J., Boddy, J., & Salvi F. (2013). *Pedagogy, curriculum, teaching practices and teacher education in developing countries*. Final Report. Education Rigorous Literature Review. Department for International Development.

Zimbabwe Ministry of Primary and Secondary Education. (2015). *Curriculum framework for primary and secondary education for period 2015–2022*. Harare: Ministry of Primary and Secondary Education.

Zimbabwe School Examinations Council. (2011a). *O-level syllabus 2011–2020: Intergrated science (5006)*. ZIMSEC.

Zimbabwe School Examinations Council. (2011b). *O level syllabus 2011–2020: Biology (5008)*. ZIMSEC.

Zimbabwe STEM Education: Policies and Practices

Abstract In this chapter, we used document analysis to examine Zimbabwe's STEM education policies and practices. We present an overview of the Zimbabwe government's initiatives, curriculum integration, teacher development, infrastructure, partnerships and inclusivity efforts. We discuss the importance of STEM education in cultivating a skilled workforce and driving socio-economic growth in the country. We also highlight challenges in implementing effective STEM education and suggest strategies to overcome them. Moreover, we evaluate the impact of STEM education on students' performance, career opportunities, and national development. We conclude the chapter by recommending further measures to advance STEM education in Zimbabwe, ensuring a stronger foundation for future generations.

Zimbabwe's STEM Education Policies and Practices

To understand Zimbabwe's STEM education policies and practices, we reviewed and interpreted curriculum documents, including syllabi. We analysed the policy documents related to STEM education in Zimbabwe. STEM policies are crucial for encouraging STEM education in Zimbabwe by providing a framework and support for developing and implementing STEM programs. These policies aim to promote the importance of STEM education, improve access to STEM resources, enhance teacher training, and create a conducive environment for STEM learning. In the next section, we present themes from the document analysis and discuss how policies encourage STEM education in Zimbabwe.

B. Chirinda et al., *STEM Education in Poverty*,
https://doi.org/10.1007/978-3-031-92490-3_5

Curriculum Development

STEM policies guide the development of STEM-focused curricula that integrate science, technology, engineering, and mathematics subjects. These curricula are designed to provide students with a comprehensive understanding of STEM concepts and foster critical thinking, problem-solving, and analytical skills. The government has implemented curriculum reforms to incorporate STEM subjects in primary and secondary education. This ensures that students are exposed to STEM concepts from an early age. For instance, the competence-based curriculum, Education 5.0, and the heritage-based curriculum.

Teacher Training and Professional Development

STEM policies emphasise the need for continuous teacher training and professional development in STEM education. They provide guidelines and resources for equipping teachers with the necessary knowledge and pedagogical skills to teach STEM subjects effectively. This ensures that teachers are well-prepared to deliver quality STEM education in classrooms. Zimbabwe provides professional development opportunities for teachers to enhance their knowledge and skills in STEM education, thereby improving the quality of STEM education in schools. For instance, the government has implemented the Science Teaching Education Programme (STEP) to improve access to STEM education in secondary schools (Mufanechiya & Makgalwa, 2024). This initiative aims to enhance the quality of STEM instruction by providing teachers with resources, training, and support, ultimately fostering students' interest and competence in STEM fields.

Infrastructure and Resource Allocation

STEM policies address the need for adequate infrastructure and resources to support STEM education. This includes providing well-equipped science laboratories, technology resources, textbooks, educational materials, and access to relevant software and tools. The policies also advocate for the equitable distribution of these resources to ensure all students have equal opportunities to engage in STEM learning. The Government of Zimbabwe, in partnership with the Zimbabwe Manpower Development Fund (ZIMDEF), is committed to strengthening STEM education as a key driver of national development and industrialisation. Through strategic funding and policy initiatives, this collaboration enhances the capacity of higher and tertiary education institutions to produce graduates with the critical skills needed in a rapidly evolving technological landscape. Under government direction, ZIMDEF actively supports infrastructure development projects at selected higher learning institutions, strengthening STEM education in Zimbabwe (Bhukuvhani et al., 2023). This investment has facilitated the construction of essential facilities such as libraries, laboratories, and engineering workshops in universities and polytechnic schools. Notable projects include the Harare Polytechnic Central Library, the Chinhoyi University of Technology (CUT) Hotel, and the Kwekwe Polytechnic Engineering Workshop.

ZIMDEF also funds groundbreaking STEM initiatives, such as commercialising the National University of Science and Technology (NUST)'s DNA Testing services and CUT's cattle breeding and artificial insemination project, contributing to scientific research and technological advancement. To enhance STEM education and training, ZIMDEF finances the procurement of advanced learning equipment, including computers, engineering materials, workshop machinery, laboratory tools, and training consumables. Additionally, state-of-the-art automotive equipment has been supplied to all polytechnics, ensuring that students are trained with industry-relevant technologies.

The government invests in the development of infrastructure, laboratories, and resources necessary for effective STEM education. This includes equipping schools with computers, internet access, and science equipment.

Collaboration and Partnerships

Zimbabwe's STEM policies encourage collaboration among educational institutions, government agencies, industry partners, and non-profit organisations. These partnerships facilitate sharing resources, expertise, and best practices in STEM education. They also promote the involvement of industry professionals and experts in mentoring and providing students with real-world experiences. The government collaborates with industries and organisations in the STEM sector to create internship programs, mentorship opportunities, and career guidance for STEM students. This helps students understand the practical applications of STEM and facilitates their transition into the workforce. For example, ZIMDEF plays a critical role in developing Zimbabwe's STEM workforce by financing the training of apprentices in STEM fields (Bhukuvhani et al., 2023). Through its funding initiatives, ZIMDEF ensures that apprentices receive hands-on, industry-relevant training in key technical and engineering disciplines. By supporting apprenticeship programs in areas such as mechanical engineering, electrical engineering, automotive technology, and industrial manufacturing, ZIMDEF helps bridge the gap between theoretical knowledge and practical application. This initiative not only enhances skills development but also strengthens Zimbabwe's industrial and technological capabilities, ensuring a steady supply of highly skilled STEM professionals for the nation's economic growth.

Scholarships and Incentives

STEM policies often include provisions for scholarships, grants, and incentives to encourage students to pursue STEM fields. These initiatives aim to attract and retain talented individuals in STEM disciplines addressing any financial barriers that students may encounter. Zimbabwe offers scholarships and grants to students pursuing STEM fields at tertiary institutions. For example, the Higher Life Foundation Joshua Nkomo Scholarship allocates 70% of the funding towards STEM studies (Tsverukayi & Poshai, 2024). This financial support encourages students to choose STEM disciplines and reduces financial educational barriers. Zimbabwe organises national STEM competitions and Olympiads to promote interest and excellence in STEM subjects. These events provide platforms for students to showcase their skills and knowledge in STEM fields.

Research and Innovation

Zimbabwe's STEM policies recognise the importance of research and innovation in STEM education. They support initiatives that promote scientific research, encourage students to engage in STEM projects and competitions, and provide funding for research and development activities in STEM fields. In line with Zimbabwe's vision to achieve middle-income status by 2030, the MHTEISTD has shifted university education from a 3.0 to a 5.0 model since 2019 (Tagwira, 2019). While the 3.0 model focused on teaching, research, and community service, the 5.0 model adds innovation and industrialisation as key components. The ministry's vision for Education 5.0 is closely tied to a heritage-based doctrine that emphase the use of local resources and cutting-edge STEM knowledge to modernise and industrialise the country (Murwira, 2019). This new approach encourages Zimbabwean universities to be innovative and creative, producing goods and services that address local challenges and contribute to the nation's development.

By implementing these various strategies, STEM policies in Zimbabwe create an environment that enables STEM education. They help promote a culture of inquiry, creativity, and problem solving among students, preparing them for future STEM careers and contributing to the country's development.

STEM Policies and Poverty

The STEM policies enacted in Zimbabwe have the potential to contribute to poverty alleviation; however, their effectiveness in achieving this goal may vary depending on various factors. By emphasising STEM education, Zimbabwe aims to equip students with the knowledge and skills needed for careers in high-demand sectors such as technology, engineering, and science (Chitate, 2016). This can increase employment opportunities for individuals, potentially leading to improved income and poverty reduction. STEM education and policies can foster technological innovation and economic growth. By producing a skilled workforce in STEM fields, Zimbabwe can attract investments, promote entrepreneurship, and create job opportunities in sectors that contribute to economic development (Svodziwa, 2021). This, in turn, can have positive effects on poverty alleviation. STEM policies need to address issues of inclusivity and ensure that education is accessible to all, including individuals from disadvantaged backgrounds. This involves providing scholarships, grants, and support systems to reduce financial barriers to STEM education. By promoting equal access to quality education, STEM policies can help address the inequalities perpetuating poverty.

STEM education can also contribute to sustainable development, crucial for poverty alleviation. By focusing on environmental sciences, renewable energy, and other sustainable practices, Zimbabwe can address pressing challenges such as climate change and resource management, ultimately benefiting marginalised communities and promoting long-term poverty reduction. While implementing STEM policies alone may not directly solve all poverty-related issues, they can significantly create opportunities, empower individuals, and foster economic growth. These policies must be complemented by broader socio-economic strategies, targeted interventions, and inclusive measures to ensure the benefits reach those most in need.

Policies and Initiatives

Document analysis revealed that Zimbabwe has adopted various policies and initiatives to promote and implement STEM education in its education system. Since gaining independence in 1980, one of the primary objectives of the Zimbabwean government, as noted by Hove and Zinyama (2012), was to develop a "scientific and technological capacity as the nerve centre for social and economic development…" (p. 285). This objective was highlighted in the government's Growth with Equity statement of 1981, which stressed the need for a comprehensive science and technology policy. However, despite references to the proposed policy in the 1982–1985 Transitional Development Plan and the First (1986–1990) and Second (1991–1995) 5-year development plans, no significant progress had been made by 2002.

The Education Act of 1987 provides the legal framework for providing education in Zimbabwe. The act states that the objectives of education are to develop the full potential of every person, to foster a sense of national identity and unity, to promote scientific and technological advancement and to prepare students for life and work in a changing society (Chitate, 2016). It is noted that from 1987 to 2000, three subjects at the secondary level, composed of mathematics, English and science, were administered without due consideration of technology and the emerging discipline of computer science.

The Education for All (EFA) policy of 1980 led to a significant increase in the number of universities offering STEM programs in Zimbabwe, growing from just one in 1952 to ten by 2019 (MHTEISTD, 2019). The University of Zimbabwe (UZ), founded in 1952 as the University College of Rhodesia and Nyasaland, reoriented its curriculum toward national development needs after independence, establishing the Faculty of Veterinary Services in 1982 (Lind, 1995). As of 2021, UZ hosts several STEM-related faculties, including Agriculture, Environment and Food Sciences, Engineering and the Built Environment, Medicine and Health Sciences, and Science (including Computer Science). Other institutions offering STEM programs were established later, such as Harare Institute of Technology (HIT) in 1988, the National University of Science and Technology (NUST) in 1990, Bindura University of Science Education (BUSE) in 1996, Chinhoyi University of Technology (CUT) in 2001, Marondera University of Agricultural Science and Technology (MUAST) in 2015, and Manicaland State University of Applied Sciences (MSUAS) in 2016.

The National Non-Formal Education Policy is rooted in a rights-based framework for providing basic education and was founded on the Universal Declaration of Human Rights (UDHR). The UDHR recognises compulsory education as a universal entitlement, while the Convention on the Rights of the Child (1989) affirms the right of all children to free and compulsory primary education. In exercising its obligation to ensure that each person enjoys the right to education, the Government of Zimbabwe has developed the following legal instruments to inform the provision of National Non-Formal Education. In 2002, Zimbabwe launched its first science and technology policy, with the primary objective, as described by Lamarchand (2014), to "…promote national scientific and technological self-reliance" (p. 89). The same year, the Department of Science and Technology was established within the Office of the President and Cabinet. Its immediate priority areas included "…rapid industrialisation; R&D focused on import substitution; fast-tracking ICTs; building human capital through science education, and popularising science and technology" (Lamarchand, 2014, p. 90). Unfortunately, the department did not appear to make any significant impact. The Zimbabwean government developed another Science and Technology Policy that led to establishing the Ministry of Science and Technology Development in 2005.

The Science, Technology and Innovation Policy 2012 outlines the vision, mission, goals and strategies for developing and promoting science, technology and innovation (STI) in Zimbabwe (Chitate, 2016). The policy aims to create a culture of innovation, enhance the quality and relevance of STI education and training, strengthen the linkages between STI and industry and increase the funding and support for STI activities. The education goal of the policy aimed to support efforts by education sector ministries by requiring them to ensure that (a) primary and secondary school students spend at least 30% of their total learning time studying science subjects (which, in practice, meant that O-level secondary students would need to take mathematics, which was compulsory, along with at least two other science subjects), (b) practical experiments are conducted in ways that leverage students' background experiences while fostering interest across genders, (c) science subjects are taught in a manner that allows each student to engage in regular hands-on experimentation, (d) both teachers and students consistently engage in research on scientific topics of interest and (e) at least 60% of university education focuses on learning science and technology skills.

The 2012 STI policy established a strong foundation for partnerships between universities and industry, which, according to Chanakira and Kanhukamwe (2014), can lead to beneficial outcomes. These partnerships facilitate the sharing of knowledge, skills, and resources for research and classroom learning, ultimately enhancing lecturers' effectiveness in teaching, research, community service, innovation, and industrialisation. Additionally, the STI policy positively influences student learning by promoting these collaborations.

The STEM initiative of 2016 was launched by the MHTESTD to encourage and support students to pursue STEM subjects and careers (Chitate, 2016). The initiative provides scholarships, mentorships, exposure to STEM role models, and opportunities for students who enrol in STEM programmes at secondary and tertiary levels. From the Ministry of Higher and Tertiary Education, Science and Technology Development perspective, the STEM initiative is structured explicitly to: (1) Increase the number of STEM students who will enrol in STEM degree programmes at the country's universities in 2018 [and beyond]; (2) Promote STEM careers in response to Zim Asset's human capital objectives; (3) Develop STEM skills that are critically needed for the country's new industrialisation thrust; (4) Address the problem of unemployment; and (5) Empower young people by promoting science and innovation (Mberi & Phambili, 2016, p. 12). The MHTESTD initiative, launched in 2016, gained significant popularity, benefiting as many as 10,445 students by 2018. The new minister of the MHTESTD discontinued the project in 2018, stating that it fell outside the ministry's mandate (Mukeredzi, 2018). Despite this, students already part of the program continued to receive its benefits. The policy aims to increase the number of students enrolled in STEM courses and support schools in improving the quality of STEM education. The three pillars of the STEM education policy are:

(a) Providing science teachers with specialised training or postgraduate qualifications in their areas of expertise to improve their teaching methods and skills.
(b) Encouraging the development and support of science, technology, and innovation centres that promote STEM learning outside schools, thus engaging the wider community.
(c) Ensuring the availability of relevant learning materials, such as textbooks, laboratory equipment, and computers.

The Curriculum Framework for Primary and Secondary Education 2015–2022 was introduced by the Ministry of Primary and Secondary Education (MPSE) to reform and update the curriculum in line with the changing needs and aspirations of students and society (Chitate, 2016). The curriculum framework emphasises the integration of STEM across all learning areas, the development of twenty-first-century skills, and the use of learner-centred and inquiry-based pedagogies. The new curriculum prioritises the teaching and learning of STEM subjects and entrepreneurship development (MoPSE, 2015). The new curriculum has introduced learning areas in agriculture, science, mathematics, and technology learning areas from the

Early Childhood Development (ECD) phase in primary school up to the 6th Form. Since the introduction of Early Childhood Development (ECD) in Zimbabwean primary schools, several policies have been enacted to ensure quality education. These include Statutory Instrument 106 of 2005, the Mathematics and Science Infant (ECD-Grade 2) Syllabus, and The Curriculum Framework for Primary and Secondary Education 2015–2022 (Mugweni, 2016). These policies aim to provide high-quality programs for all preschool children.

The current reforms shaping school science in Zimbabwe—Science, Technology, Engineering, and Mathematics (STEM), STEM and Arts (STEAM) education, and Heritage-Based Education 5.0 (HBE 5.0) are powerful tools for developing a range of competencies (knowledge, skills, values, and attitudes) that equip students to thrive in twenty-first-century economies (MHTESTD, 2018; Ministry of Primary and Secondary Education [MoPSE], 2015). STEM education policies have become a critical focus for countries worldwide due to the high demand for skilled professionals in these fields. Zimbabwe recognises this demand and has developed policies to promote STEM education.

In 2021, the Zimbabwean government launched another STEM education programme called the STEM Junior City Programme. The programme targets children aged between 6 and 14 years old and seeks to inspire and promote an interest in STEM subjects, aiming to improve STEM skills from a young age. To complement established policies, in 2020, the president of Zimbabwe launched the Presidential STEM Scholarship, targeting the top 100 students from all of Zimbabwe's ten provinces. This scholarship covers tuition, accommodation, and other necessary expenses for outstanding students to follow STEM courses at one of the country's four science & technology state universities.

Challenges and Gaps in the Implementation of STEM Education

Despite the policies and initiatives, Zimbabwe still faces many challenges and gaps in implementing STEM education in its schools and colleges. Some of these challenges and gaps are:

- There is a shortage of adequately qualified STEM teachers, especially in rural and remote areas. Many teacher education programmes in Zimbabwe do not adequately prepare pre-service teachers to teach STEM subjects and integrate STEM practices in their classrooms (Mashoko & Veal, 2023). The study also found that many in-service teachers lack the content knowledge, pedagogical skills, and confidence to teach STEM effectively
- There is a lack of sufficient and appropriate STEM resources, such as textbooks, laboratories, equipment, and technology. Many schools and colleges in Zimbabwe do not have access to the basic and essential resources needed to support STEM teaching and learning. This limits the opportunities for students to engage in

hands-on and practical activities, experiments and projects that foster STEM interest, and understanding.

- There is a lack of alignment and coherence between the STEM policies and the curriculum (Mpofu, 2019). Although the STEM policies and the curriculum framework share similar goals and objectives, a gap exists between the policy formulation and curriculum implementation. For instance, the STEM initiative mainly focuses on the secondary and tertiary levels, whereas the curriculum framework encompasses both the primary and secondary levels. There is also a lack of coordination and collaboration among the different ministries and stakeholders involved in STEM education, such as the MHTESTD, the MPSE, the universities, the industry, and the civil society.
- There is a lack of awareness and appreciation of the value and relevance of STEM education among students, parents, teachers, and society (Gadzirayi et al., 2016). Many students and parents in Zimbabwe have negative perceptions and attitudes towards STEM subjects and careers due to the subjects' difficulty, complexity, competitiveness, and lack of role models and guidance. Many teachers and society do not recognise the importance and benefits of STEM education for the personal, social, and economic development of the students and the country.

Actions and Recommendations

To address challenges and gaps, Zimbabwe must adopt a holistic and systematic approach to implementing STEM education within its education system. To improve both the quality and quantity of STEM teachers, Zimbabwe should invest more in the professional development and support of pre-service and in-service teachers. This includes equipping them with adequate and relevant content knowledge, pedagogical skills, and resources to effectively teach STEM subjects and integrate STEM practices into their classrooms. Additionally, it involves creating opportunities for them to network, collaborate, and learn from other STEM teachers, experts, and mentors. To enhance the availability and accessibility of STEM resources, Zimbabwe must mobilise additional funding and support from the government, private sector, donors, and the community to provide and maintain the necessary and appropriate resources for STEM teaching and learning. This encompasses textbooks, laboratories, equipment, technology, and online platforms. It also involves developing and sharing locally relevant and contextualised STEM materials and activities that reflect the needs and interests of students and society.

In Zimbabwe, the Ministry of Primary and Secondary Education and the Ministry of Higher and Tertiary Education, Science and Technology Development have differing views on the definition of STEM education and its implications for implementation (Yingi et al., 2022). Consequently, Zimbabwe still lacks a clear and accessible national STEM education framework (Mpofu, 2019). Stakeholders need to reach a consensus on a unified understanding of STEM education and its implementation in educational settings (Mpofu, 2019). This chapter argues that to overcome the persistent challenges of STEM education reform, academics must analyse and synthesise various theoretical frameworks into practical, easy-to-understand, and implementable approaches. Zimbabwe must establish and strengthen coordination and collaboration among the ministries and stakeholders involved in STEM education to ensure alignment and coherence between STEM policies and the curriculum. This includes developing and implementing a common vision, mission, goals, and strategies for STEM education and monitoring and evaluating the progress and impact of STEM policies and initiatives. It also involves engaging students, teachers, parents, and society in policy formulation and the curriculum development processes.

To enhance the awareness and appreciation of the value and relevance of STEM education among students, parents, teachers, and society, the government can create and share positive and inspiring messages, stories, and images related to STEM subjects and careers. This effort should also celebrate and showcase the achievements and contributions of STEM students and professionals. Additionally, this endeavour can include providing and promoting greater access to STEM role models and opportunities such as clubs, camps, fairs, competitions, and scholarships.

References

Bhukuvhani, C., Nezamndonyi, J., & Mutanga, P. (2023). Advancement of Science, Technology and Innovation (STI) education in Zimbabwe: Policies and practices (Chapter 2). In B. Chirinda, L. Sibanda, J. Vere, & G. Sunzuma (Eds.), *Science, Mathematics and Technology education in Zimbabwe* (pp. 27–44). Peter Lang. issn 2576-358x (print), issn-3598 (online).

Chanakira, M., & Kanhukamwe, O. C. (2014). Role of universities in contributing towards science and technology park development: A framework of critical success factors. In D. S. Oh & F. Phillips (Eds.), *Technopolis, best practices for science and technology cities* (pp. 299–311). Springer.

Chitate, H. (2016). Science, Technology, Engineering and Mathematics (STEM): A case study of Zimbabwe's educational approach to industrialisation. *World Journal of Education, 6*(5), 27–35.

Gadzirayi, C. T., Bongo, P. P., Ruyimbe, B., Bhukuvhani, C., & Mucheri, T. (2016). *Diagnostic study on the status of STEM in Zimbabwe:* Collaborative research by the Bindura University of Science Education and higher life foundation, Technical Report (pp. 1–47). Bindura: BUSE.

Hove, M., & Zinyama, T. (2012). The challenges of Zimbabwe science and technology policy formulation from 1980 to 2002. *International Journal of Humanities and Social Science, 2*(2), 284–291.

Lamarchand, G. A. (2014). *Mapping research and innovation in the Republic of Zimbabwe.* UNESCO. Retrieved from https://www.academia.edu/13104480/Mapping_Research_and_Innovationin_the_Republic_of_Zimbabwe?pop_sutd=false

Lind, A. (1995). *Desk study of the education sector in Zimbabwe.* Education division document No. 66. Swedish International Development Agency.

Mashoko, D., & Veal, W. R. (2023). Implementing STEM policy in African nations' teacher education programs: Insights from some Southern African countries. In S. M. Al-Balushi, L. Martin-Hansen, & Y. Song (Eds.), *Reforming science teacher education programs in the STEM era* (Palgrave studies on leadership and learning in teacher education). Palgrave Macmillan. https://doi.org/10.1007/978-3-031-27334-6_2

Mberi, N., & Phambili, M. (2016, 28 February–5 March, 12). Science, Technology, Engineering and Mathematics craze hits Zimbabwe. *The Sunday News*. STEM supplement.

Ministry of Higher and Tertiary Education, Science and Technology Development. (2019). *Strategic plan 2019–2023*. Zimbabwe Ministry of Higher and Tertiary Education, Science and Technology Development. Retrieved from mhtestd.gov.zw

Ministry of Higher and Tertiary Education, Science and Technology Development [MHTESTD]. (2018). *Doctrine: The modernisation and industrialisation of through education, science and technology development*. Zimbabwe Government.

Ministry of Primary and Secondary Education. (2015). *Curriculum framework for primary and secondary education 2015–2022 (CBC 2015–2022)*. Government Printers.

Mpofu, V. (2019). A theoretical framework for implementing STEM education. In G. Fomunyam (Ed.), *Theorizing STEM education in the 21st century*. IntechOpen. https://doi.org/10.5772/intechopen.88304

Mufanechiya, A., & Makgalwa, M. M. (2024). Re-visioning Zimbabwe's Science, Technology, Engineering and Mathematics (STEM) for secondary schools through the science teacher education programme. *Journal of Education and Learning Technology, 5*(6), 111–129. https://doi.org/10.38159/jelt.2024564

Mugweni, R. M. (2016). Early childhood (ECD) teachers' conceptions and implementation of the child-centred approach to teaching science. *International Journal of Academic Research in Progressive Education and Development, 5*(4), 2226–6348.

Mukeredzi, T. (2018). Furore as the government stops popular STEM programme. *University World News*. Retrieved from https://www.universityworldnews.com/post.php?story=20180126063955233

Murwira, A. (2019). *Towards revitalising the roles of universities in development [Zimbabwe]*. Ruforum.

Svodziwa, M. (2021). *STEM education as a human capital development strategy for the industrialization and modernization of Zimbabwe*. Retrieved from SSRN: https://ssrn.com/abstract=3788852 or https://doi.org/10.2139/ssrn.3788852

Tagwira, F. (2019). *Strategic plan 2019–2023*. Zimbabwe Ministry of Higher and Tertiary Education, Science and Technology Development.

Tsverukayi, B. H., & Poshai, L. (2024). Commodification of Zimbabwe's higher education amid funding decline: Impact on universal access. *Transformation in Higher Education, 9*, a399. https://doi.org/10.4102/the.v9i0.399

Yingi, E., Hlungwani, P., & Nyagadza, B. (2022). The Fourth Industrial Revolution (4IR) in the heart of the SDG agenda: The role of education in Zimbabwe. *Africa Review, 14*(2022), 213–229.

Access and Equity in STEM Education in Zimbabwe

Abstract Using a desk review, in this chapter, we provide an overview of the historical context of STEM education in Zimbabwe, highlighting the challenges, opportunities, and interventions to promote access and equity. The Zimbabwean education system has implemented several reforms. One of the major factors driving such reforms was a focus on improving STEM education through increasing practical learning and transitioning from a content-based curriculum to a competency-based one. The aim was to encourage students to continue with STEM subjects. Hence, STEM education is offered at various academic levels, ranging from ECD to higher and tertiary education. Additionally, students from underrepresented groups (e.g., those from rural educational institutions and females) were encouraged to participate in STEM education. Even though Zimbabwe has devoted resources to address various STEM education-related challenges, research has revealed a lack of priority towards consolidating and improving access, quality, and quantity in STEM pedagogy. The inspiration behind emphasising STEM education was to prepare the twenty-first-century workforce by exposing students to related classroom activities that would enable them to solve real-world situations and become competitive in this era of technology. However, the Zimbabwean competency-based curriculum which embraces arts and science education discriminates through STEM streaming, posing challenges to twenty-first-century trends of inclusive education for all. Given such a challenge, this chapter dwells on access and equity in STEM Education in Zimbabwe.

Introduction

In recent years, the global push for advancements in STEM has underscored the importance of equitable access to quality education. Zimbabwe, like many other nations, recognises the pivotal role STEM education plays in fostering innovation, economic development, and societal progress. As a highly literate society, Zimbabwe benefits significantly from tapping into the opportunities presented by the STEM sector. Scholars and experts have emphasised that leveraging human-centered

B. Chirinda et al., *STEM Education in Poverty*,
https://doi.org/10.1007/978-3-031-92490-3_6

innovations within STEM fields could potentially uplift the country from poverty, inadequate service delivery, and sluggish industrialisation (Makuyana & Mbulayi, 2019; Dekeza & Kufakunesu, 2017). Makuyana and Mbulayi (2019) highlight that STEM education offers valuable skills highly sought after in the job market, thereby addressing the issue of unemployment. Similarly, Ngara (2017) underscores the high status and job satisfaction associated with STEM careers and their potential to offer competitive salaries and social prestige. Moreover, Ngara (2017) points out the global trend where STEM fields are recognised for their capacity to drive economic growth, with projections indicating significant potential for stimulating growth rates of 15–20% points in the coming decade. This projection is particularly promising for African economies, like Zimbabwe, given their highly literate populations that can readily engage in STEM-related activities and foster collaborations to support the production and consumption of STEM products and services.

Human capital emerges as a critical catalyst among the various factors contributing to economic growth and increased productivity. Effective investment in STEM education is identified as a key component in realising Zimbabwe's industrialisation and modernisation goals (Svodziwa, 2021). STEM education is pivotal in fostering innovation, technological advancement, and skill development, all essential for driving economic growth and industrial transformation. By equipping individuals with the necessary knowledge and expertise in STEM fields, Zimbabwe can enhance its capacity for innovation, attract investment in high-tech industries, and compete more effectively in the global marketplace.

Furthermore, STEM education empowers individuals with the skills and competencies needed to adapt to the demands of a rapidly evolving labor market characterised by technological change and automation. By investing in STEM education, Zimbabwe can cultivate a highly skilled workforce capable of driving innovation, increasing productivity, and sustaining long-term economic growth. Furthermore, STEM experts argue that STEM careers are forward-looking and offer favorable economic prospects. Makuyana and Mbulayi (2019) found that STEM graduates face lower unemployment risks than their peers in other fields post-graduation.

Investing in STEM education and fostering a conducive environment for STEM-related innovation and employment could significantly contribute to Zimbabwe's socio-economic development, offering pathways out of poverty, enhancing industrialisation, and empowering its highly literate populace. The burgeoning interest among socioeconomic development stakeholders in Zimbabwe in adopting STEM as a pivotal driver for progress is a significant development (Makuyana & Mbulayi, 2019). This chapter uses a desk review to examine the disparities within the country's STEM education system that have emerged with the newfound focus on STEM.

Historical Context

The historical context of STEM education access and equity in Zimbabwe is deeply rooted in the nation's historical and sociopolitical developments (Chitate, 2016). Key historical periods provide insights into the evolution of access and equity in STEM education in Zimbabwe.

Pre-Independence Era (Before 1980): Educational Inequities and STEM Disparities in Colonial Zimbabwe

The pre-independence era in Zimbabwe, marked by colonial rule, was characterized by profound disparities in educational opportunities, particularly in STEM education (Chitate, 2016). During the colonial period, educational opportunities were skewed in favor of the white minority population, and access to STEM education was particularly limited for the majority black population. The colonial authorities established an educational system that primarily served the interests of the white minority population, neglecting the majority black population (Zindi, 1996). The curriculum emphasised skills needed for manual labor rather than fostering scientific and technological advancements (Chitate, 2016). Educational institutions were segregated, with few resources allocated to schools serving black students, contributing to disparities in STEM education access (Mudaly & Chirikure, 2023).

During this period, educational institutions were segregated based on racial lines, with schools for white students receiving more resources, qualified teachers, and advanced curricula. In contrast, schools for black students were often underfunded, lacked essential resources, and operated with a curriculum that emphasised basic skills suitable for manual labor rather than intellectual pursuits (Chitate, 2016). The intentional educational disparities perpetuated a hierarchical society, where the white minority had access to quality education and opportunities, while the black majority faced systemic discrimination and limited access to educational advancements, including STEM education (Mudaly & Chirikure, 2023).

Access to STEM education during the colonial era was particularly restricted for the black population. STEM subjects were often considered a privilege reserved for the white elite, contributing to a significant knowledge gap between the two racial groups. Black students were not provided the necessary resources, laboratories, or equipment for a comprehensive STEM education. The curriculum itself was tailored to meet the needs of the colonial administration, focusing on producing a workforce skilled in agriculture and basic manual labor rather than fostering scientific inquiry or technological advancements (Zindi, 1996; Kanyongo, 2005). Consequently, STEM education was marginalised and underdeveloped, limiting the intellectual growth and potential of the majority black population (Mudaly & Chirikure, 2023).

In Zimbabwe, segregation extended beyond the classroom to resource allocation (Zindi, 1996). Black schools lacked essential infrastructure, including well-equipped laboratories and libraries, which hindered effective STEM instruction. The lack of qualified teachers in these schools further compounded the challenges faced by black students interested in pursuing STEM fields (Kanyongo, 2005). The deliberate neglect of black educational institutions, coupled with the restricted access to STEM education, perpetuated a cycle of inequality, limiting the social mobility and opportunities available to the majority black population (Kanyongo, 2005).

The pre-independence era in Zimbabwe was characterised by stark educational disparities that significantly impacted access to STEM education. The colonial legacy of intentional segregation, resource imbalances, and a curriculum geared towards manual labor shaped the foundations of educational inequality (Zindi, 1996). The post-independence period would bring about efforts to dismantle these barriers and establish a more inclusive educational system to address historical injustices and promote equity in STEM education.

Post-Independence Transformation (1980s–1990s): Democratizing STEM Education in Zimbabwe

The post-independence reforms focused on eliminating historical racial discrimination and expanding educational opportunities for Africans (Kanyongo, 2005; Chitate, 2016). The post-independence era in Zimbabwe, beginning in the 1980s, marked a pivotal period during which the newly established government undertook comprehensive efforts to address historical educational inequalities, specifically focusing on STEM education. The overarching goal was to democratise access to education and make STEM education more inclusive, bridging the gaps entrenched during the colonial era. Policies were implemented to democratise access to education, emphasising the inclusion of STEM education (Mudaly & Chirikure, 2023). Initiatives were undertaken to establish new schools and improve infrastructure in previously marginalised areas (Ndlovu, 2013). The government also aimed to increase the number of black students pursuing STEM subjects, addressing historical imbalances.

Following independence in 1980, the Zimbabwean government, led by Prime Minister Robert Mugabe, recognised the need for transformative changes in the education system. Policies were implemented to ensure that all citizens, regardless of race or socioeconomic background, had equal opportunities to access quality education, including STEM subjects. Initiatives such as the Education Act of 1987 laid the foundation for an education system that aimed to be more inclusive and responsive to the entire population's needs (Kanyongo, 2005). This legislation was crucial in reshaping the educational landscape and promoting equity, diversity, and accessibility.

Quality, access, and relevance issues are vital in scaling up STEM education across different African groups (Fomunyam, 2020). One of the key strategies to democratise STEM education in Zimbabwe involved the establishment of new schools and the improvement of infrastructure, particularly in areas that had been marginalized during the colonial period. The Zimbabwean government invested in building schools and upgrading existing facilities in rural and underserved urban areas, ensuring that students in these regions had access to the same quality of education as those in more affluent areas (Kanyongo, 2005). This expansion of educational infrastructure included the development of science laboratories, computer rooms, and libraries equipped with resources necessary for STEM education. The goal was to create an environment conducive to effective teaching and learning in STEM subjects, breaking away from the historical neglect of these areas.

To address the historical imbalances in STEM education, the government implemented programs aimed at increasing the participation of black students in STEM subjects. Scholarships and financial assistance programs were introduced to support students from disadvantaged backgrounds in pursuing higher education in STEM fields (Ngwenya et al., 2017). Additionally, teacher training programs were initiated to enhance their skills, ensuring a cadre of qualified teachers capable of delivering quality STEM education. This approach was crucial in nurturing the next generation of scientists, engineers, and technologists from diverse backgrounds.

Recognising the importance of cultural relevance in education, the government took steps to make the curriculum more inclusive and reflective of the local context (Chitate, 2016). Efforts were made to integrate indigenous knowledge systems into STEM education, making the subjects more relatable and engaging for students from different cultural backgrounds (Chigwedere, 2007). Language inclusivity was also moved towards teaching STEM subjects in indigenous languages alongside English. This shift aimed to eliminate language barriers and ensure students had a more accessible learning experience in STEM. Culturally responsive teaching acknowledges and respects students' cultural backgrounds, connecting STEM education to their lived experiences. Integrating indigenous knowledge systems and incorporating local examples in STEM lessons makes the content more relatable. This approach helps students see the relevance of STEM in their communities, fostering a sense of belonging and promoting equity in access.

The post-independence transformation in Zimbabwe during the 1980s and 1990s was characterised by a commitment to addressing historical inequalities in STEM education. The government laid the groundwork for a more inclusive and equitable education system through policy initiatives, infrastructure development, and targeted programs. The emphasis on democratising access to STEM education set the stage for cultivating a diverse and skilled workforce, contributing to the nation's aspirations for development and progress in the STEM fields.

Despite efforts to encourage participation in STEM fields among all students, a disparity persists in the distribution of STEM degrees earned, with underrepresented groups not holding an equitable share. The post-independence government aimed to achieve equality of educational opportunity, often referred to as democratisation (Ndlovu, 2013). However, despite efforts, neither equality nor equal opportunity was achieved. Instead, the deep racial divide inherited from the colonial era widened, creating a broader gap between social classes and between urban and rural children (Ndlovu, 2013).

Economic Disparities in STEM Education in Zimbabwe

Economic barriers present significant challenges for students from lower-income backgrounds in accessing quality STEM education. These barriers often stem from disparities in school resources, including inadequate funding for advanced laboratory equipment and a shortage of experienced teachers. Research by Vadvel et al. (2023) highlights the impact of socioeconomic factors on STEM education. In Zimbabwe, schools in economically disadvantaged areas often lack the resources necessary to offer hands-on experiences and access to state-of-the-art equipment crucial for effective STEM learning (Marongedza et al., 2022; Mwembe et al., 2022). Additionally, budget constraints can result in larger class sizes and a shortage of qualified STEM teachers, further exacerbating the problem (Svodziwa, 2021; Marongedza et al., 2022). These disparities disadvantage students from lower socioeconomic backgrounds, limiting their exposure to STEM concepts, hands-on experimentation, and opportunities for STEM-related extracurricular activities. As a result, students from wealthier backgrounds may have a competitive advantage when applying to colleges or pursuing careers in STEM fields.

Poverty can impact the quality of teaching in STEM subjects, particularly if teachers in disadvantaged areas lack adequate training and professional development opportunities (Svodziwa, 2021; Mwembe et al., 2022). Economic disparities also influence the quality of teaching staff and professional development opportunities for teachers (Marongedza et al., 2022). Affluent urban schools can attract and retain qualified STEM teachers through better remuneration and professional development opportunities (Marongedza et al., 2022). Conversely, rural schools may struggle to secure and retain qualified STEM teachers which can impact the overall quality of instruction (Svodziwa, 2021; Marongedza et al., 2022). Economic disparities in STEM education in Zimbabwe perpetuate a cycle of inequality, limiting opportunities for students in economically disadvantaged areas.

Mabhanda (2016) highlights that the prospects of STEM education in Zimbabwe are facing significant challenges due to the deteriorating economic conditions. This economic downturn has resulted in many educational institutions struggling to establish and sustain the necessary infrastructure for STEM education and research (Gadzirayi et al., 2016). Consequently, the development and advancement of STEM fields in Zimbabwe are hindered by inadequate resources and facilities, ultimately impacting the nation's ability to compete globally in science and technology (Gadzirayi et al., 2016).

Geographical Disparities in STEM Education in Zimbabwe

Geographical disparities in Zimbabwe have long been a significant factor influencing access to quality education, particularly STEM education. The urban-rural divide has resulted in distinct challenges hindering effective STEM education in rural areas, where limited infrastructure, resources, and qualified teachers contribute to a stark contrast in learning experiences (Mwembe et al., 2022; Nera & Nyikadzino, 2023). Geographical limitations pose significant challenges for students in rural or remote areas, resulting in disparities in access to quality STEM education and resources compared to their urban counterparts. Research by Nera and Nyikadzino (2023) underscores the impact of geographic location on educational opportunities, particularly in STEM fields.

Poverty often results in unequal resource distribution, disproportionately affecting schools in economically disadvantaged areas (Svodziwa, 2021; Marongedza et al., 2022; Mwembe et al., 2022). The majority of Zimbabwe's population resides in rural areas, where statistics indicate that 5311 primary schools (84.46%) and 2264 secondary schools (78.86%) are situated. In rural or remote areas, schools may face constraints such as limited funding, a shortage of qualified teachers, and insufficient infrastructure for STEM education (Nera & Nyikadzino, 2023; Marongedza et al., 2022; Svodziwa, 2021). Rural school infrastructure in Zimbabwe hinders the realisation of Sustainable Development Goal (SDG) number 4 by 2030, a United Nations target to ensure inclusive and equitable education and promote lifelong learning opportunities for all. The poor education infrastructure, particularly in the approximately 700 satellite schools on former commercial farms following the land reform program, poses a significant obstacle to achieving this goal. This can lead to a lack of access to advanced laboratory equipment, specialised courses, and extra-curricular STEM activities. Technological advancements are integral to modern STEM education, yet access to technology is unevenly distributed (Chirume & Kwabena, 2016). With improved computer power, faster data transfer rates, and lower costs, even developing countries like Zimbabwe have seen a surge in Internet use (Chirume & Kwabena, 2016). By January 2014, about 40% (5.2 million) of Zimbabwe's population were Internet users, compared to 15.7% in 2011 and 0.4% in 2000 (Kabweza, 2014). Despite this rapid increase, research indicates that e-learning adoption in areas such as primary and secondary schools has not reached its full potential in Zimbabwe (Chitanana et al., 2008). There appears to be an uneven distribution of ICT use for teaching and learning across secondary and primary schools and the workforce (Chirume & Kwabena, 2016). For example, students in rural areas often have limited access to computers and the Internet (Chirume & Kwabena, 2016). While this issue is not unique to Zimbabwe, it contradicts the government policy on providing quality education through ICT for all (Chirume & Kwabena, 2016). Additionally, the digital divide hampers Zimbabwe's commitment to the Education for All movement, a global pledge to provide quality basic education for all children, youth, and adults (EFA Global Monitoring Report, 2013/2014). With better financial standing, urban schools can afford computers, internet connectivity, and other technological resources that enhance the learning experience (Marongedza et al., 2022). The digital divide exacerbates existing inequalities in an era where technology plays a crucial role in STEM education. Students from impoverished backgrounds may lack access to computers, the internet, and other technological tools that enhance STEM learning. This digital divide deprives them of the benefits of online resources, educational software, and interactive learning materials that enhance the overall STEM learning experience. As a result, students in rural or remote areas may miss out on valuable opportunities to explore STEM subjects, engage in hands-on learning experiences, and pursue STEM-related career paths (Mwembe et al., 2022). This perpetuates inequalities in educational outcomes and limits the potential for economic and social advancement in these communities.

Laboratories play a critical role in STEM education, providing students with opportunities to experiment, observe, and apply theoretical concepts (Bhukuvhani et al., 2012). However, many rural schools lack adequately equipped laboratories or may not have them at all (Gudyanga & Kurup, 2017; Svodziwa, 2021; Mwembe et al., 2022). The absence of such facilities limits students' exposure to the practical aspects of STEM subjects, impeding their ability to develop essential skills and gain a deep understanding of scientific principles.

The shortage of qualified STEM teachers is another pressing issue in rural areas (Svodziwa, 2021; Kadziya & Ndebele, 2020; Marongedza et al., 2022). Many experienced teachers prefer urban schools due to their better amenities, professional development opportunities, and improved living conditions (Marongedza et al., 2022). As a result, rural schools may have fewer teachers with specialised knowledge in STEM subjects (Kadziya & Ndebele, 2020). The lack of qualified teachers hampers the delivery of quality STEM education, leaving students with limited guidance and mentorship (Kadziya & Ndebele, 2020; Mabhanda, 2016; Mukomana, 2019; Kadziya & Ndebele, 2020; Mwembe et al., 2022).

Geographical disparities also manifest in transportation challenges (Marongedza et al., 2022). Rural schools may be located far from urban centers or lack reliable public transportation, making it difficult for students to access supplementary resources, attend extracurricular activities, or participate in STEM-related events and competitions (Dekeza & Kufakunesu, 2017). Affluent urban schools often have the financial means to offer a variety of extracurricular STEM activities, such as science clubs, competitions, and field trips (Dekeza & Kufakunesu, 2017. These activities contribute to a more holistic and engaging STEM learning experience. In contrast, economic constraints in rural areas limit such opportunities, depriving students of valuable hands-on experiences and exposure to real-world applications of STEM concepts (Dekeza & Kufakunesu, 2017).

Gender Disparities in STEM Education in Zimbabwe

The gender gap in STEM fields is a global issue, with disparities in both developed and developing countries. However, the severity of the problem appears to be more pronounced in developing countries, including those in the African region, and Zimbabwe is not an exception (Matete, 2022). Throughout Zimbabwe history, gender dynamics in STEM education have mirrored global trends, characterized by persistent disparities and evolving efforts to promote gender equity. Gender disparities in STEM education persist in Zimbabwe, reflecting deeply ingrained societal norms and stereotypes that often discourage girls from pursuing STEM subjects (Gudyanga & Kurup, 2017; Gweshe & Chiware, 2023). Studies, such as those cited by Makuyana and Mbulayi (2019), emphasize that underrepresented minority groups and females are disproportionately represented in STEM degrees at both the undergraduate and graduate levels. The impact of gender biases is reflected in the underrepresentation of women in STEM fields at higher education institutions and in professional settings (Maisiri, 2024). Fewer women pursuing STEM careers contribute to a lack of diversity in these fields, limiting the range of perspectives and ideas (Maisiri, 2024). This underrepresentation extends beyond academia into the workforce, creating an equity gap in career opportunities and advancement for women in STEM (Maisiri, 2024).

Women remain significantly underrepresented among STEM graduates in Zimbabwe, with female students comprising only 19% of the total compared to 39% of male students (Chingarande et al., 2023). Research on this disparity has identified structural factors within social institutions and labor market segmentation as key contributors (Gudyanga, 2016). These factors are rooted in deeply entrenched patriarchal and cultural norms and practices that limit women's agency and reinforce their subordinate status (Chingarande et al., 2023; Zvavahera & Chirima, 2023). A study by Gudyanga et al. (2015) found that gender norms within Zimbabwean culture, which prioritise men over women, influenced girls' perceptions of science and math as "masculine and difficult subjects." This perception often conflicted with their cultural definition of a woman, negatively impacting their participation in these subjects (Gudyanga et al., 2015).

The participation of women in STEM fields is further hindered by negative stereotypes and a lack of role models, which lowers girls' performance and aspirations in science and technology (Chingarande et al., 2023). Role models play a crucial role in boosting confidence and raising the aspirations of female students (Vurayai, 2012). Additionally, covert discrimination, implicit biases, and societal expectations regarding appropriate roles and careers for women contribute to the gap (Gudyanga, 2016). These societal constraints often force women to make choices that steer them away from STEM disciplines (Chingarande et al., 2023).

Traditional gender roles, societal expectations, and deep-rooted stereotypes have historically discouraged girls from pursuing STEM subjects in Zimbabwe, creating challenges for female students aspiring to enter these fields (Gudyanga & Kurup, 2017). In Zimbabwe, Social norms often dictate that certain fields, particularly those in STEM, are more suitable for men, while women are encouraged to pursue careers in non-STEM fields (Gudyanga & Kurup, 2017). These stereotypes reinforce the idea that STEM subjects require inherent abilities more commonly associated with males, creating a barrier for girls interested in these areas. Reflecting these biases, educational institutions provided fewer opportunities and resources for female students in STEM. These biases contribute to the underrepresentation of women in STEM fields, creating a notable equity gap in educational and career opportunities.

For instance, STEM streaming, as described in Zimbabwean schools, categorises students into different academic tracks or classes based on their perceived aptitude or performance in science and mathematics subjects (Mukomana, 2019). This practice often leads to the segregation of students into "fast-learner" and "slow-learner" categories, with the former being encouraged to focus more on STEM subjects while the latter may face barriers to accessing quality STEM education. The implications of STEM streaming are multifaceted and can have significant consequences for individual students and the broader educational system. Students categorised as "slow learners" may be denied access to advanced STEM classes or resources, perpetuating educational inequality. This limits their opportunities for academic and career advancement in STEM fields. Being labeled as a "slow learner" based on performance in specific subjects can negatively affect students, leading to diminished self-esteem and motivation. This stigma may also affect how teachers perceive and interact with these students, exacerbating the problem. STEM streaming restricts students' future career options by directing them towards or away from STEM subjects based solely on their performance in these areas. This can reinforce gender and socioeconomic disparities in STEM fields, as streaming practices may disproportionately affect certain groups.

In addition, traditional gender roles and societal expectations in Zimbabwe may perpetuate stereotypes that discourage girls, especially those from impoverished backgrounds, from pursuing STEM subjects (Gudyanga & Kurup, 2017). Chirimuuta et al. (2012) found that tradition and culture in many African homes often require girls to be always available to assist their mothers with household chores, leaving them with inadequate time to study STEM subjects. Girls from impoverished backgrounds may face additional barriers to accessing quality education, including STEM subjects (Gudyanga & Kurup, 2017). Economic constraints can result in limited resources for educational materials, transportation, and even basic necessities such as uniforms. As a result, girls in poverty may experience a compounded disadvantage in pursuing STEM education. The intersection of gender and poverty can intensify these challenges, reinforcing stereotypes that limit girls' educational aspirations in STEM. Research in Zimbabwe has indicated that teachers require additional training in gender responsiveness, as well as the need for positive female role models in STEM classrooms (Tikly et al., 2018).

Studies, such as Makuyana and Mbulayi (2019) and Makarova et al. (2019), highlight the persistent, wide gender gap in STEM disciplines in sub-Saharan African countries. The persistence of gender disparities in STEM education creates an equity gap that limits the educational opportunities available to girls. Fewer girls pursuing STEM subjects at the secondary and tertiary levels results in a skewed representation in STEM-related courses and disciplines (Chingarande et al., 2023). This affects individual career trajectories and impacts the nation's overall capacity to harness diverse talent in driving innovation and scientific advancements. Women consistently lag behind men in STEM education and employment opportunities related to science fields (Chingarande et al., 2023). UNESCO (2017) and Mukhwana et al. (2020) also emphasise the underrepresentation of women in STEM industries, with women scientists being outnumbered by men. Various factors contribute to this gap, including insecurity, lack of self-efficacy, and a perceived lack of support or representation for females in STEM fields.

Women, particularly girls in rural schools, may feel insecure in choosing STEM-related majors due to a lack of female role models in STEM and a potentially unwelcoming educational environment (Gudyanga & Kurup, 2017). Mandiudza (2021) and Gudyanga and Kurup (2017) also suggest that the gender gap in STEM education may stem from a lack of confidence or support for females in these fields. The scarcity of female role models and mentors in STEM, particularly among those from impoverished backgrounds, can influnce girls' perceptions of their potential in these fields (Gudyanga & Kurup, 2017). The absence of relatable role models may hinder girls' ability to envision themselves succeeding in STEM careers, perpetuating a cycle of underrepresentation (Gadzirayi et al., 2016).

Mandiudza (2021) found that males tend to have higher self-efficacy in STEM-related activities during middle school, which influences their likelihood of choosing STEM courses and careers. In contrast, females exhibit lower self-efficacy, making them less likely to pursue STEM paths. Gudyanga and Kurup (2017) further support this notion by demonstrating that females enrolled in STEM subjects, such as Physics, may have lower confidence levels than their male counterparts, potentially leading to a higher likelihood of females leaving STEM majors. According to Zindi (1996), even those girls who go further to higher schools end up opting for training courses in nursing and teaching after the Ordinary level ('O' Level) due to lower aspirations. Researchers such as Matete (2022) have identified these disparities and their underlying causes. Women and minorities may face barriers in accessing quality STEM education due to systemic inequalities in resources, educational opportunities, and support networks. Additionally, societal stereotypes and biases can discourage individuals from these groups from pursuing STEM careers or lead to perceptions of unwelcomeness within STEM fields.

Unfortunately, several challenges have been identified as obstacles to the involvement of rural girls in the STEM discourse. According to Masanja (2010), rural areas face a significant challenge in terms of low participation by female students and teachers in STEM subjects. The limited number of female students excelling in mathematics at the Ordinary level significantly affects the number of those pursuing STEM subjects at the Advanced level and later at the degree level (Mandiudza, 2021). According to Dekeza and Kufakunesu (2017), despite the Zimbabwean government's efforts to sponsor students interested in pursuing STEM subjects in high school, very few girls have been able to take advantage of these opportunities.

Societal norms and expectations often discourage females from considering careers in STEM fields, perpetuating the belief that these areas are more suitable for males (Dekeza & Kufakunesu, 2017).

Financial constraints force many young girls to drop out of school, limiting their access to STEM education and opportunities (Ngara, 2017; Isaacs et al., 2023). The literature suggests that while economic hardships affect all members of society, females often experience more pronounced effects due to poverty, leading to exclusionary outcomes and limited personal development opportunities in STEM education (Gudyanga & Kurup, 2017). Zimbabwean parents tend to prioritise the education of their sons over that of their daughters at the secondary level when resources are limited, and they often encourage young women to get married after completing primary school (Zindi, 1996). To date, this trend has not been phased out completely in Zimbabwe. In 1990, only half of the girls of secondary school age were enrolled, compared to three-quarters of the boys (Zindi, 1996). That same year, the government reinstated school fees, increasing the dropout rate among girls (Zindi, 1996). By 1994, girls made up just 42% of secondary school enrollments, despite the efforts of NGOs and donor agencies like CIDA, to promote girls' education (Zindi, 1996). In 1995, the University of Zimbabwe implemented an affirmative action policy that allowed girls with lower entry qualifications than boys to enroll in degree courses after 'A' Levels (Zindi, 1996; Chingarande et al., 2023). However, despite these initiatives, less than 40% of the enrolled students were girls. Gudyanga and Kurup (2017) argue that this situation perpetuates the subjugation of women.

The prevalence of teenage pregnancy and early marriages further diminishes girls' chances of pursuing STEM education and careers (Ngara, 2017; Isaacs et al., 2023). Girls are often steered towards traditional feminine careers like nursing and teaching instead of being encouraged to explore their potential in STEM fields (Gudyanga & Kurup, 2017). Moreover, Makuyana and Mbulayi (2019) highlight additional challenges faced by young girls in STEM, such as menstrual periods, caregiving responsibilities for sick family members, impassable rivers during rainy seasons, and participation in food-for-work schemes with their families, which contribute to girls missing school and thus opportunities to engage in STEM education. These multifaceted challenges collectively hinder the participation and success of rural girls in STEM, perpetuating the gender gap in these fields and limiting the potential talent pool for scientific and technological advancement in Zimbabwe.

Students with Disabilities and STEM Education

The underrepresentation of students with disabilities in STEM majors and STEM-related careers is a significant issue that can be attributed to various factors, including systemic barriers and a lack of support. In Zimbabwe, students with disabilities may encounter physical, sensory, or cognitive barriers that make it difficult to access STEM learning environments, resources, and equipment (Muranda-Kaseke & Mbawuya, 2021). These barriers can hinder their ability to fully participate in

STEM coursework and activities without adequate accommodations and support services.

According to Isaacs et al. (2023), in Zimbabwe, academic institutions may not have comprehensive policies and practices to support the needs of students with disabilities in STEM fields. For example a study conducted in Zimbabwe by Matorevhu (2023) showed that the Ministry of Primary and Secondary Education (MoPSE)'s policy document, that is the Curriculum Framework for Primary and Secondary Education 2015–2022, that there is no reference to how visual impaired students should be involved in ICT integration as a teaching and learning tool in STEM education. This includes the unavailability of assistive technologies, lack of accessible instructional materials, and faculty training on inclusive teaching methods for students and teachers.

In Zimbabwe, there were also limited resources in the form of assistive devices such as Braille machines and other visual technologies for the blind, hearing aids and cochlear implants for the deaf, computers for those with intellectual challenges or learning disabilities, and mobility devices such as wheelchairs for those with physical disabilities (Sibanda, 2018; Dube et al., 2021). Furthermore, the lack of ICT hardware and software posed a significant challenge to implementing STEM education in Zimbabwe. All children with disabilities can benefit from using ICT (Sibanda, 2018). ICT-enabled pedagogical and assistive devices are particularly beneficial for children with disabilities (Sibanda, 2018). For instance, software such as JAWS is crucial for including blind students. Unfortunately, such technologies remain scarce in Zimbabwe (Sibanda, 2018). Poor or inadequate funding is the primary reason for the lack of these and other resources. Generally, the lack of funding is mainly responsible for the inaccessibility of schools, especially those in rural areas (Sibanda, 2018).

In Zimbabwe, students with disabilities may perceive STEM fields as unwelcoming or hostile environments due to stigma, stereotypes, and misconceptions about disability (Majoko, 2019; Dube et al., 2021). This can lead to feelings of isolation, discrimination, and lack of belonging, ultimately discouraging students with disabilities from pursuing STEM majors or careers (Majoko, 2019). Students with disabilities often lack visible role models and mentorship opportunities in STEM fields, making it challenging for them to envision themselves succeeding in these areas (Matorevhu, 2023). Without positive representation and support from peers and professionals with disabilities, students with disabilities may feel discouraged or marginalised.

Students with disabilities may face additional financial and resource constraints that impact their ability to pursue STEM education and career opportunities (Matorevhu, 2023). This includes costs associated with assistive technologies, specialised support services, accommodations, and limited access to scholarships and financial aid tailored to students with disabilities in STEM fields (Sibanda, 2018).

Researchers have explored various strategies to increase the participation of underrepresented groups in STEM-related college majors and careers. Mandiudza (2021) in Zimbabwe proposed implementing educational tours, STEM quiz competitions, and STEM Olympiads to broaden students' understanding and interest in

STEM professions. Nkala et al. (2024) proposed service learning projects and community involvement with engineering activities as effective methods to cultivate an interest in and positive attitude towards STEM among underrepresented groups.

Conclusion

In Zimbabwe, as in many developing nations, the quest for access and equity in STEM education is a journey fraught with challenges yet brimming with opportunities for transformative change. Throughout this exploration, we have delved into the intricacies of Zimbabwe's educational landscape, identifying barriers that hinder equitable access to STEM opportunities while also acknowledging the strides made towards inclusivity.

Achieving access and equity in STEM education is essential for fostering innovation, economic development, and social progress in Zimbabwe. By addressing systemic barriers, promoting inclusivity, and investing in educational infrastructure, Zimbabwe can unlock the potential of its youth and cultivate a diverse and skilled workforce capable of driving sustainable growth in the twenty-first century. Stakeholders must collaborate effectively to ensure that every Zimbabwean student has equal opportunities to pursue and excel in STEM education, contributing to the nation's prosperity and global competitiveness.

References

Bhukuvhani, C., Mupa, M., Mhishi, M., & Dziva, D. (2012). Science practical work instructional technologies and open distance learning in science teacher training: A case study in Zimbabwe. *International Journal of Education and Development Using ICT, 8*(2), 17–27.

Chigwedere, E. (2007). *Cultural policy of Zimbabwe*. Ministry of Education, Art, Sport and Culture.

Chingarande, S. D., Muchabaiwa, W., & Muyambo, T. (2023). *Bridging the gender equality gap in science at the women's university in Africa, Zimbabwe*. Final Technical Report submitted to IDRC.

Chirimuuta, C., Bhukuvhani, C., & Gudhlanga, E. (2012). Towards a gender inclusive curriculum in Zimbabwe's education system: Opportunities and challenges. *Gender and Behaviour, 10*(1), 4533–4545.

Chirume, E., & Kwabena, D. (2016). Toward a better understanding of the access to education through online learning in the case of Zimbabwe. *International Journal of Advanced Research, 4*(7), 1304–1313. https://doi.org/10.21474/ijar01/1031

Chitanana, L., Makaza, D., & Madzima, K. (2008). The current state of e-learning at universities in Zimbabwe: Opportunities and challenges. *International Journal of Education and Development Using Information and Communication Technology, 4*(2), 5–15.

Chitate, H. (2016). Science, Technology, Engineering and Mathematics (STEM): A case study of Zimbabwe's educational approach to industrialisation. *World Journal of Education, 6*(5), 25–35.

Dekeza, C., & Kufakunesu, M. (2017). Implementation of STEM curriculum in rural secondary schools in Zimbabwe: Limits and possibilities. *Journal of Emerging Trends in Educational Research and Policy Studies, 8*(1), 11–15.

Dube, T., Ncube, S. B., Mapuvire, C. C., Ndlovu, S., Ncube, C., & Mlotshwa, S. (2021). Interventions to reduce the exclusion of children with disabilities from education: A Zimbabwean perspective from the field. *Cogent Social Societies, 7*(1), 1–18. https://doi.org/10.1080/23311886.2021.1913848

EFA Global Monitoring Report. (2013/2014). *Teaching and learning: Achieving quality for all*. UNESCO.

Fomunyam, K. G. (2020). Agenda 2063: Science, Technology, Engineering, and Mathematics (STEM) education to the rescue. *International Journal of Engineering Research Technology, 13*, 1603–1609. https://doi.org/10.37624/IJERT/13.7.2020.1603-1609

Gadzirayi, C. T., Bongo, P. P., Ruyimbe, B., Bhukuvhani, C., & Mucheri, T. (2016). *Diagnostic study on the status of STEM in Zimbabwe*. Bindura University of Science Education and Higher Life Foundation.

Gudyanga, A. (2016). Zimbabwean female participation in physics: Factors of identity formation considered as contributing to developing an orientation to physics by female students. *Journal of Education and Practice, 5*(26), 159–171.

Gudyanga, A., & Kurup, R. (2017). Zimbabwean female participation in physics: The influence of identity formation on perception and participation. *Cogent Education, 4*(1), 1–8.

Gudyanga, A., Adam, K., & Kurup, R. (2015). Zimbabwean female participation in physics: The influence of context on identity formation. *African Journal of Research in Mathematics, Science and Technology Education, 19*(2), 172–184. https://doi.org/10.1080/10288457.2015.1050805

Gweshe, G. T., & Chiware, M. (2023). Gender-inclusive education in Science, Technology, Engineering, and Mathematics (STEM) fields in post-independence Zimbabwe. In D. Seck, J. U. Elu, & Y. Nyarko (Eds.), *Advances in African economic, social and political development* (Vol. Part F1230, pp. 295–310). https://doi.org/10.1007/978-3-031-31431-5_16

Isaacs, N., Middleton, L., Essop, R., Lynch, I., Fluks, L., Chikwanha, M., Chinyemba, F., Gumede, Z., Djoukouo, F., Kuetche, I., Maseko, N., Ndinda, C., Tamangani, R., Tendenedzai, J., Agugua, A., & Van Rooyen, H. (2023). *Strengthening gender equality and inclusivity in the national system of Science, Technology and Innovation: Zimbabwe country profile*. HSRC.

Kabweza, L. S. M. (2014, October 7). Here's a list of Africa's video on demand & pay TV providers. TechZim. https://www.techzim.co.zw/2014/10/heres-a-list-of-africasvideo-on-demand-pay-tv-providers/

Kadziya, L., & Ndebele, C. (2020). An analysis of the challenges faced by teachers in the delivery of Science, Technology, Engineering and Mathematics education in rural day secondary schools in Zimbabwe. *IAR Journal of Engineering and Technology, 1*(3), 85–95.

Kanyongo, Y. G. (2005). Zimbabwe's public education system reforms: Successes and challenges. *International Education Journal, 6*(1), 65–74.

Mabhanda, W. (2016). Opportunities and factors affecting adoption of STEM education: The case of Gweru Polytechnic first year commerce students. *International Journal of Business Marketing and Management, 1*(5), 18.

Maisiri, J. (2024). Preparing Zimbabwean women for the future of work: The role of STEM education in the so-called fourth industrial revolution. *South African Review of Sociology*, 1–18. https://doi.org/10.1080/21528586.2024.2311399

Majoko, T. (2019). Inclusion of children with disabilities in physical education in Zimbabwean primary schools. *Sage Open, 9*(1). https://doi.org/10.1177/2158244018820387

Makarova, E., Aeschlimann, B., & Herzog, W. (2019). The gender gap in STEM fields: The impact of the gender stereotype of math and science on secondary students' career aspirations. *Frontiers of Education, 4*(60), 1–11. https://doi.org/10.3389/feduc.2019.00060

Makuyana, A., & Mbulayi, S. P. (2019, July 8–11). *Beyond the enclave: Expediting the challenges and prospects of young girls doing STEM subjects in public high schools in rural Zimbabwe*. Being a paper presented at Al-Azhar International Conference Center Cairo, Egypt

by Conference Of Rectors, Vice Chancellors & Presidents of Association African Universities. Theme: Role of Higher Education Institutions in Promoting the Continental Education Strategy for Africa (CESA 16–25).
Mandiudza, M. (2021). Factors motivating female students from Masvingo, Zimbabwe—The choice of taking natural sciences at university. *European Journal of Social Sciences Studies, 6*(2). https://doi.org/10.46827/ejsss.v6i2.1001
Marongedza, L., Hlungwani, P. M., & Hove, P. (2022). Institutional constraints affecting secondary school student performance: A case study of rural communities in Zimbabwe. *Cogent Education, 10*(1). https://doi.org/10.1080/2331186X.2022.2163552
Masanja, V. G. (2010). *Increasing women's participation in science, mathematics and technology education and employment in Africa*. Paper presented at the United Nations Division for the Advancement of Women (DAW, part of UN Women) & United Nations Educational, Scientific and Cultural Organization (UNESCO): Expert group, meeting Gender, Science and Technology, Paris: Special project on Scientific, Technical and Vocational Education of Girls in Africa.
Matete, R. E. (2022). Why are women under-represented in STEM in higher education in Tanzania? *FIRE: Forum for International Research in Education, 7*(2), 48–63. https://doi.org/10.32865/fire202172261
Matorevhu, A. (2023). Curriculum innovation implementation for industrialization: A case of education 5.0 preservice science and mathematics teacher preparation. *Journal of Research in Instructional, 3*, 69–86. https://doi.org/10.30862/jri.v3i1.214
Mudaly, R., & Chirikure, T. (2023). STEM education in the Global North and Global South: Competition, conformity, and convenient collaborations. *Frontiers of Education, 8*(1144399), 1–13. https://doi.org/10.3389/feduc.2023.1144399
Mukhwana, A., Abuya, T., Matanda, D., Omumbo, J., & Mabuka, J. (2020). *Factors which contribute to or inhibit women in Science, Technology, Engineering, and Mathematics in Africa*. The African Academy of Sciences (AAS).
Mukomana, S. (2019). Teaching challenges affecting stem resuscitation in Zimbabwe. *European Journal of Education Studies, 5*(11), 321–332.
Muranda-Kaseke, K., & Mbawuya, K. D. (2021). Leveraging empowerment for those students with disabilities: Towards creating an inclusive university education framework. *International Journal of Research and Innovation in Social Science, 5*(2), 594–601.
Mwembe, D., Zulu, I. L., & Mdlongwa, P. (2022). Needs assessment for facilitators in teaching of STEM subjects in under resourced rural schools: Case of Lupane and Hwange Districts in Zimbabwe. *International Journal of Professional Development, Learners and Learning, 4*(1), ep2202. https://doi.org/10.30935/ijpdll/11666
Ndlovu, M. E. (2013). *Zimbabwe's educational legacy from the 1980s: Was it all so rosy?* Kubatana.net.
Nera, G. R., & Nyikadzino, T. (2023). Factors hampering the realization of equity and quality education in Zimbabwe's rural schools: Experiences of Chipinge Central Circuit, Manicaland. *Journal of Asian and African Studies*. https://doi.org/10.1177/00219096231207892
Ngara, C. (2017). Gifted education in Zimbabwe. *Cogent Education, 4*(1), 1332840.
Ngwenya, B., Pelser, T., & Sibanda, R. (2017). Success as the knowledge economy the success or failure of the Zimbabwe's STEM initiative. In *International conference on intellectual capital and knowledge management and organisational learning* (pp. 169–178). Academic Conferences International.
Nkala, B., Sibanda, V. M., Ndhlovu, J., Hobane, L., & Singh, S. (2024). Challenges and drivers of industrialist propensity among chemical engineering students in STEM institution in Zimbabwe: Towards a conceptual framework. *International Journal of Environment, Engineering and Education, 6*(1), 26–37. https://doi.org/10.55151/ijeedu.v6i1.126
Sibanda, P. (2018). A review of the implementation of inclusive education in Zimbabwe: Challenges and opportunities. *Scientific Journal of Pure and Applied Sciences, 7*(9), 808–815.

Svodziwa, M. (2021, February 19). *STEM education as a human capital development strategy for the industrialization and modernization of Zimbabwe*. Retrieved from SSRN: https://ssrn.com/abstract=3788852 or https://doi.org/10.2139/ssrn.3788852

Tikly, L., Joubert, M., Barrett, A. M., Bainton, D., Cameron, L., & Doyle, H. (2018). *Supporting secondary school STEM education for sustainable development in Africa*. Bristol working papers in education. Bristol: School of Education, University of Bristol.

UNESCO. (2017). *UNESCO and gender equality in Sub-Saharan Africa: Innovative program. Visible results*. UNESCO.

Vadivel, B., Alam, S., Nikpoo, I. & Ajanil, B. (2023). The impact of low socioeconomic background on a child's educational achievements. *Education Research International*. https://doi.org/10.1155/2023/6565088

Vurayai, S. (2012). *Equity principle in mathematics education: Focus on ordinary mathematics*. Lambert Academic.

Zindi, F. (1996). Towards the elimination of disparities in educational provision: A look at Zimbabwe and South Africa. *Journal of Social Development in Africa, 11*(1), 43–51.

Zvavahera, P., & Chirima, M. (2023). Flexible work arrangements and gender differences in research during the COVID-19 period in Zimbabwean higher learning institutions. *Perspectives in Education, 41*(1), 4–18.

Teachers' Perspectives on the Effectiveness of STEM Education in Zimbabwe

Abstract With technological advancements, global efforts are being made to develop high-quality STEM education. STEM education is an instructional approach that integrates science, technology, engineering, and mathematics by connecting them to everyday situations. Through STEM education, Zimbabwean students are expected to acquire knowledge in diverse fields and gain twenty-first-century skills such as creativity, critical thinking, and problem-solving, making them competitive in the global market. STEM education is implemented from preschool to higher education within formal and informal educational settings. Consequently, teachers are crucial for effectively adapting STEM education and addressing the challenges encountered. Teachers' perspectives and their preparedness influence the success of STEM education implementation; hence, it is vital for them to receive adequate training and to understand STEM education, which would assist in executing STEM-related activities in the classroom. For this reason, it is of utmost importance to comprehend how STEM teachers perceive STEM education, as this directly impacts their teaching and learning methods in the classroom. This chapter focuses on Zimbabwean teachers' perspectives on the effectiveness of implementing and integrating STEM education in Zimbabwean schools. We hope the findings from the study can assist educational institutions globally in designing teacher empowerment courses, planning future education policies, and understanding teachers' needs to improve STEM education.

Introduction

The role of teachers in STEM education extends beyond instruction; they are key drivers of innovation and curriculum adaptation (Madusise, 2020). Their perspectives provide valuable insights into the practical challenges and opportunities associated with STEM teaching and learning. By examining teachers' views, policymakers and educators can develop more effective strategies to support STEM education and enhance student engagement. As education systems evolve to meet

B. Chirinda et al., *STEM Education in Poverty*,
https://doi.org/10.1007/978-3-031-92490-3_7

the demands of a rapidly changing world, understanding teachers' perspectives on the effectiveness of STEM education becomes essential. As key stakeholders, teachers often face additional burdens in adapting to new teaching methodologies and incorporating technology into their classrooms. Understanding teachers' perspectives is crucial for gauging the success and sustainability of STEM education initiatives. Teachers are instrumental in designing and delivering STEM curricula that align with the dynamic needs of students and the workforce (Moyo & Berdut, 2017). Successfully implementing STEM education relies heavily on teachers' dedication, expertise, and innovation. As frontline facilitators, teachers play a central role in shaping the STEM curriculum, translating its principles into meaningful student learning experiences. Teachers provide unique insights into the challenges they face, their strategies, and the impact of STEM education on students. In underdeveloped countries like Zimbabwe, where the education system is under strain, teachers' perspectives shed light on the potential transformative power of STEM education.

STEM Education

The successful implementation of STEM education is heavily influenced by various teacher-related factors, including their perspectives (Manunure & Leung, 2024). Teachers' perspectives play a crucial role in shaping the effectiveness of STEM education, as they determine how concepts are introduced, contextualised, and integrated into students' learning experiences. Their beliefs, attitudes, and pedagogical approaches influence how students engage with STEM subjects, particularly in making abstract concepts more accessible and meaningful. Teachers who recognise the relevance of STEM education beyond the classroom are more likely to implement interdisciplinary and inquiry-based learning strategies that bridge theoretical knowledge with practical applications. Furthermore, their insights into students' diverse backgrounds, learning styles, and prior knowledge help create inclusive and culturally responsive STEM learning environments that enhance engagement and achievement.

From this perspective, STEM education links the teaching and learning of STEM concepts to real-world experiences, providing students with hands-on, relevant learning opportunities. It equips students to apply mathematical skills to solve real-life problems using engineering and design principles. STEM education emphasises the study of mathematics and science while meaningfully integrating technology and engineering design. This integration fosters innovative problem-solving, makes mathematics and science more relevant and engaging, and helps students develop the critical skills and career connections necessary to thrive in an economy that lacks skilled workers and problem solvers.

STEM education is a methodology that encourages students to explore and address issues that are pertinent to the world they live in (Madusise, 2020). Classroom experiences should mirror real-world scenarios, providing students with meaningful opportunities to engage in problem-solving. STEM education is designed to cultivate students who are problem solvers, innovators, inventors, self-reliant thinkers, logical thinkers, and technologically literate individuals (Morrison, 2006). This multidisciplinary approach integrates science, technology, engineering, and mathematics to foster real-world applications and prepare students for the demands of the twenty-first-century workforce (Madusise, 2020). However, despite the potential benefits of STEM education, its effective implementation presents significant challenges, particularly in the context of Africa. For instance, a study by Kadziya and Ndebele (2020) highlighted several challenges that hinder the advancement of STEM education across Africa, including insufficient supervision strategies, inadequate scientific infrastructure, a shortage of qualified teachers, and a lack of essential teaching and learning materials. Additionally, limitations in pedagogical content knowledge and curriculum deficiencies stemming from inadequate quality assurance during the design stage further impede effective STEM education (Kadziya & Ndebele, 2020).

Research, such as that by Mutseekwa (2021), has revealed gaps in the structured integration of STEM education within the Science Teacher Education curriculum, with teachers identifying deficiencies in STEM-related competencies. This lack of integration has hindered efforts to promote STEM literacy and preparedness among students. Furthermore, Dekeza and Kufakunesu (2017) highlighted the challenges faced by rural secondary schools, where teachers reported insufficient resources, inadequate teacher support, and a lack of proper infrastructure such as laboratories. These issues have led to the failure of rural schools to fully implement STEM curricula, thereby limiting the potential for effective STEM education delivery.

Given that teachers are the primary facilitators of STEM education in classrooms, it is crucial to understand their perspectives and experiences regarding the challenges and opportunities associated with STEM education. Teachers' perspectives are essential to addressing the gaps in the implementation process and improving the delivery of STEM education. This study, therefore, seeks to explore Zimbabwean teachers' perspectives on STEM education, to identify the factors that influence their ability to successfully teach STEM subjects and to provide recommendations for enhancing the integration and effectiveness of STEM education in Zimbabwean schools. The following research question guides the study:

What are teachers' perspectives on the effectiveness of implementing and integrating STEM education in Zimbabwean schools?

Methodology

The theoretical framework underpinning this study was social constructivism, which posits that individuals construct their understanding based on the interplay between their prior knowledge, beliefs, and new ideas or experiences they encounter (Richardson, 1997). Social constructivists emphasise the role of social interaction in shaping meaning within communities (Creswell, 2015). In qualitative research, meaning is derived from participants' perspectives and interpreted by researchers (Creswell, 2015). This study employed open-ended questions (see Appendix) in focus group discussions, allowing participants to articulate their perspectives on STEM education.

Research Design

A qualitative case study design was adopted to facilitate an in-depth exploration of a phenomenon within its real-life context through multiple data sources (Baxter & Jack, 2010). Since the study investigated teachers' perspectives on STEM education, various qualitative data collection methods were utilised. The qualitative case study approach was deemed appropriate as it enabled the researchers to provide rich, contextualised descriptions of teachers' perspectives on STEM teaching and learning.

Data Collection Instruments

Focus group discussions were employed to comprehensively explore the research topic (Gray, 2011). A focus group discussion is an interactive research method in which a facilitator guides a small, homogeneous group of 6–12 participants in a structured debate on a research topic (Christensen et al., 2015). This study held five focus group discussions, each comprising eight participants (five male and three female). The study utilised open-ended questions, allowing participants to express their thoughts in full rather than selecting predefined responses. A key advantage of open-ended questions is their ability to generate rich and unanticipated responses, offering deeper insights into the research problem (Sarantakos, 2013).

Additionally, focus group discussions were conducted to gather diverse perspectives on STEM education. This approach allowed participants to express their perspectives comfortably within a group, fostering interaction and discussion. Focus group discussions are valued for their ability to generate rich qualitative data by capturing contextualised insights and encouraging participants to expand on their perspectives through interaction (Curtis & Curtis, 2011). The group setting encouraged participants to reflect on and refine their perspectives collaboratively.

Validity and Reliability

In this study, member checking was utilised to verify the accuracy of data obtained from focus group discussions (Creswell, 2015). Transcripts of these discussions were printed and provided to the in-service teachers, allowing them to review the recorded information for accuracy and fairness. Where necessary, amendments were made to ensure that the interpretations accurately reflected participants' perspectives (Creswell, 2015). This process helped establish the plausibility and trustworthiness of the study's findings.

Cultural Context of the Study

The research was conducted at a Zimbabwe university specialising in training mathematics, chemistry, physics, biology and computer science teachers. The study specifically targeted undergraduate in-service teachers enrolled in a 2-year degree program within the Faculty of Science Education. These teachers came from diverse cultural backgrounds across various provinces in Zimbabwe. Their teaching experiences in secondary schools exposed them to students from multicultural communities, reflecting the highly diverse societal contexts in which they work. This cultural diversity played a crucial role in shaping the teachers' perspectives on STEM education in Zimbabwe.

Research Participants

To explore teachers' perspectives on STEM education, participants were selected from a pool of potential informants expected to provide relevant insights (Silverman, 2011). The study focused on in-service second-year Bachelor of Science Education Honours teachers specialising in STEM subjects. These teachers were admitted to the university with a minimum teaching qualification, either a diploma or certificate in education obtained from a teacher training college or university, in addition to at least 2 years of teaching experience. The selection of in-service teachers was based on their foundational knowledge in the teaching and learning of STEM subjects. Since they were trained to teach these subjects up to the Ordinary Level ('O' Level), they possessed relevant expertise for the study. At the time of the research, there were 90 eligible teachers (55 males and 35 females). A proportional stratified random sampling technique was employed to ensure fair representation in the sample. A total of 40 in-service teachers (25 males and 15 females) were selected, ensuring that each participant had an equal chance of being included in the study. According to Seale (2012), three to five focus groups are sufficient for a study, while an ideal focus group comprises six to eight participants (Hennink, 2014). In this study, five focus groups were formed, each consisting of eight in-service teachers (five males and three females). A group of this size was deemed appropriate for facilitating in-depth discussions while allowing the moderator to gather diverse perspectives. Participants for the focus group discussions were selected randomly, ensuring the sessions fostered small-group interaction and meaningful engagement with the research topic.

Research Procedure and Data Collection

Permission to conduct the study was obtained from the university where the research was carried out. In-service teachers provided informed consent by signing consent forms, which outlined the purpose of the study, their voluntary participation, their right to withdraw at any stage, and the measures taken to ensure confidentiality. To maintain confidentiality, the five groups were assigned the codes A, B, C, D, and E. Focus group discussions were scheduled at times convenient for the participants to encourage full engagement. Each discussion session lasted approximately 1 h, ensuring sufficient time for participants to share their perspectives in depth.

Data Analysis

The qualitative data analysis in this study involved systematically selecting, sorting, categorising, synthesising, summarising, and interpreting data from focus group discussions. Various qualitative analysis techniques exist (Adams & Lawrence, 2015); however, this study employed inductive data analysis procedures to allow patterns and themes to emerge naturally from the data (Creswell, 2015). Focus group discussions were audio-recorded and transcribed, ensuring accurate data capture. Inductive analysis involves identifying patterns, categories, and themes within the data rather than imposing a fixed framework at the outset (Curtis & Curtis, 2011). For the focus group discussions, a coding framework was developed after data collection (Cohen et al., 2015). Data were systematically coded and gradually condensed into key themes and sub-themes.

Results

The research findings presented stem from an analysis of focus group discussions. The themes that emerged include teachers' conceptualisation of STEM education, the benefits of STEM education, teacher preparedness to implement STEM education, the rural and urban STEM divide, resources in STEM education, community support and engagement, and gender issues in STEM education.

Teachers' Conceptualisation of STEM Education

The effectiveness of STEM education largely depends on how teachers conceptualise and implement it in their classrooms. In this study, teachers held varying perspectives of STEM education. STEM education as defined by the in-service teachers.

> STEM education is defined as a program that enhances the understanding of knowledge by emphasising the interrelationships between different subjects and utilising appropriate learning strategies essential for solving real-life problems. (Group A)
>
> STEM education focuses on the teaching and learning of science subjects while equipping students with essential scientific knowledge. It aims to develop their professional competencies in STEM concepts, ensuring they acquire the skills necessary for future life. (Group B)
>
> STEM education is an approach that connects various fields, science, technology, engineering, and mathematics, to accomplish tasks or activities aligned with curriculum learning outcomes. STEM education is characterised as an integrated activity designed to make teaching and learning more effective and meaningful. (Group C)
>
> What we understand is that science subjects are taught separately, each with its distinct methods. There is minimal overlap between them, and the way they are learned lacks integration or meaningful connections. (Group D)
>
> STEM education is a holistic approach that fosters critical thinking, problem-solving, and innovation by encouraging students to apply knowledge from multiple disciplines practically and meaningfully. It emphasises hands-on experiences, inquiry-based learning, and real-world applications to develop students' ability to think creatively and adapt to evolving technological advancements. (Group E)

The conceptions presented reveal a range of perspectives on STEM education, from a traditional, subject-specific view (Group D) to a more integrated, holistic approach (Group E). While some teachers recognise STEM as an interdisciplinary framework (Groups A, C, E), others remain focused on science as the central component (Group B). The foundations of STEM education rest on a rich tapestry of philosophical principles, an interdisciplinary approach, hands-on learning, real-world applications, technology integration, and the cultivation of twenty-first-century skills (Tsvara et al., 2018). STEM transcends traditional boundaries as an educational paradigm, reflecting a holistic vision that prepares students for academic success and active participation in a rapidly changing society (Tsvara et al., 2018).

Benefits of STEM Education

Teachers highlighted the numerous benefits of STEM education. Teachers believe that STEM subjects are essential for developing students' knowledge and skills, enabling them to effectively address and solve real-world problems.

> STEM has the potential to greatly enhance student motivation. Students can become more independent, critical thinkers, and problem-solvers by integrating science, technology, engineering, and mathematics. This integration encourages curiosity, logical thinking, and collaboration. Additionally, STEM education is not just about gaining subject-specific

> knowledge; it fosters broader 21st-century skills, such as technological literacy and teamwork, which are essential in today's world. (Group A)
>
> STEM education has the ability to engage students in a way that traditional methods may not. We can help students develop critical thinking, problem-solving, and collaborative skills by integrating these fields. It's about more than just learning content; it's about stimulating curiosity, encouraging students to ask questions, and guiding them to find solutions independently. These skills are crucial for success in the modern world and will help prepare them for a wide range of careers. (Group B)
>
> Integrating STEM fields can motivate students to become active researchers and independent thinkers. By connecting science, technology, engineering, and mathematics, students will not only learn specific content but will also develop essential skills like critical thinking, collaboration, and technological proficiency. (Group C)
>
> STEM ensures that the knowledge acquired in the classroom is applied practically in real-life situations. (Group D)
>
> STEM education encourages students to explore new information, solve problems, and work together across disciplines, ultimately preparing them for future challenges in an ever-evolving world. (Group E)

The perspectives from these five groups highlight the multidimensional benefits of STEM education. Groups A, B, and C emphasise the importance of integrating STEM subjects to foster critical thinking, problem-solving, and collaboration. These groups view STEM as a tool for developing students' independence and essential twenty-first-century skills such as technological literacy, which are crucial for success in a modern, rapidly evolving world. Group D focuses on the practical application of classroom knowledge, emphasising that STEM education prepares students for real-world challenges. The findings are in line with Mutseekwa (2021), who noted that creativity, innovation, problem-solving, and construction of real knowledge through practical and real-time activities were crucial for STEM education. Finally, Group E highlights the exploratory nature of STEM, stressing the importance of cross-disciplinary collaboration and preparing students to navigate future challenges. Gadzirayi et al. (2016) emphasise the interconnectedness of STEM subjects, highlighting the symbiotic relationship between science, technology, engineering, and mathematics. The teachers' perspectives demonstrate that STEM education is about content acquisition and preparing students with the skills, mindset, and practical knowledge needed to succeed in an interconnected and dynamic global society (Madusise, 2020).

Teacher Preparedness to Implement STEM Education

Teacher preparedness to implement STEM education is a critical issue. Some teachers believed they were unprepared due to the lack of targeted training and practical opportunities to implement STEM education, citing gaps in their confidence and expertise to engage students fully in these subjects. This is what they said:

> Yes, many of us majored in Mathematics or Science in college, but with the emphasis on STEM in the new curriculum, we believe we should receive in-service training on how to teach STEM effectively. (Group C)

> We read about STEM in print and electronic media but have never participated in a workshop on the topic. As a result, we don't feel confident implementing the STEM curriculum. (Group E)

Teachers stated that most of them majored in Mathematics or Science in college rather than the integration of STEM subjects. The finding aligns with Manunure and Leung (2024), who believed that most in-service teachers in Zimbabwe were trained in either mathematics or science, with limited or no experience implementing integrated STEM pedagogy. The perspectives from Group C reflect a proactive approach, acknowledging the need for ongoing professional development to stay relevant in teaching STEM. Teachers in this group understand the value of hands-on, targeted training to complement their academic backgrounds. Conversely, Group E highlights a lack of direct engagement with STEM training programs, pointing to a significant gap in teacher preparedness. The findings align with those of Kadziya and Ndebele (2020), who highlighted the shortage of professional programmes that guide teachers on the connection between STEM disciplines and how to teach them in class. The teachers feel underprepared to implement the curriculum effectively, emphasising the need for workshops and practical training to boost their confidence and teaching effectiveness in STEM subjects. In line with findings, Mwembe et al. (2022) highlight the need for tailored professional development programs that address the specific challenges of STEM education, ensuring that teachers are well-equipped to deliver impactful lessons and foster a culture of lifelong learning.

Teachers expressed the need for ongoing professional development and support to effectively teach integrated STEM concepts. This is what they said:

> Ongoing professional development for STEM teachers is crucial. We need opportunities to learn new teaching methods, stay current on content, and collaborate with colleagues. When teachers feel supported and empowered, it translates to better student outcomes. (Group A)
>
> By adopting cutting-edge and captivating teaching strategies, we can make STEM subjects more approachable and enjoyable for our rural students. This is made possible by increased teacher training and support. (Group B)
>
> Investing in the professional development and ongoing support of STEM teachers is crucial. Providing teachers with robust training, resources, and collaborative opportunities can help them better engage students, adapt their instructional practices, and foster a love for STEM subjects. (Group C)
>
> It is essential for teachers to receive ongoing training, resources, and support to stay up to date with the latest developments in STEM subjects. This enables them to create engaging, interactive lessons that cater to diverse learning styles and abilities. By prioritising teacher preparedness, we can empower students to become the next generation of innovators, thinkers and problem-solvers. (Group D)
>
> We should also explore ways to attract and retain high-quality STEM teachers, particularly in our area. Offering incentives can help build a strong, dedicated STEM teaching workforce that can positively impact student learning and achievement. (Group E)

All five groups recognise the crucial role of professional development in improving the quality of STEM education. Group A, Group C, and Group D specifically highlight the need for teachers to have access to continuous training and resources to stay updated and improve their teaching practices. The findings are in line with Mutseekwa (2021), who highlighted the need for adequate training for STEM teachers if STEM implementation is to be realised in Zimbabwe. Groups B and E address more specific needs, such as making STEM subjects accessible to rural students and attracting quality teachers to underserved areas. The finding concurs with Mandiudza (2021), whose research revealed that schools in rural areas failed to attract qualified teachers. The consensus across the groups is that robust teacher training, support, and resources are essential for the successful implementation of STEM education, ensuring that students benefit from innovative and effective teaching practices.

Rural and Urban STEM Divide

The rural-urban divide in STEM education presents significant challenges for students in rural areas, where limited access to resources, expertise, and extracurricular activities hinders their learning and career development opportunities. Teachers said:

> Being situated in a rural area, we face unique challenges in accessing STEM resources and expertise. It's harder for us to bring in guest speakers, organise field trips, or collaborate with urban STEM institutions. (Group B)
>
> Inadequate funding for STEM programs is a major concern. We see that our students' peers in urban areas have access to state-of-the-art STEM facilities and resources, which puts our students at a significant disadvantage. (Group C)
>
> The lack of STEM-focused extracurricular activities and career exposure in our community makes it difficult for our students to envision themselves in STEM-related professions. This urban-rural gap needs to be addressed. (Group D)
>
> Many rural schools lack the essential infrastructure required to effectively implement STEM education. Additionally, students in these schools often lack the cultural capital needed to fully appreciate the significance of STEM education, in contrast to their peers in urban areas. In rural areas, access to resources and opportunities can be more limited. (Group E)

The responses from the five groups collectively emphasise the rural-urban divide in STEM education, with rural areas facing unique challenges in accessing resources, funding, and extracurricular opportunities. Zhou (2017) found similar results where there was a gap in resources and STEM subject uptake in schools located in rural areas. These barriers limit the ability of rural students to fully engage with and benefit from STEM education. It is important to address issues such as funding, resource allocation, and leveraging technology to provide equitable access to high-quality STEM education for all students, regardless of location. Teachers had this to say:

> Leveraging technology, virtual learning, and community outreach can help bridge that gap and ensure equitable access to high-quality STEM education. (Group C)
>
> Improved transportation and access to rural secondary schools, which would enable their students to visit well-equipped STEM laboratories and engage with STEM professionals in urban centers, would be a game changer as this would enable our students to gain self-assurance and practical skills in STEM subjects to improve their competitiveness in their job market. (Group D)

The results from the two groups propose viable solutions to the challenges posed by the rural-urban divide in STEM education. Group C advocates for technological innovation, while Group D focuses on tangible transportation and resource access improvements. These approaches suggest combining virtual and physical solutions, such as technology integration and enhanced mobility, which can significantly help rural students access high-quality STEM education and close the gap with their urban peers. In line with these findings, Mpofu and Mutseekwa (2023) believed that teachers in the digital literacy era need adequate training and skills in technology. Ruparanganda (2019) also express concerns about access to technology and the need for adequate training to harness its full potential.

Resources in STEM Education

Resources play a critical role in STEM education, as they directly influence the effectiveness of teaching and learning. Teachers had this to say:

> Our school's lack of high-quality STEM resources and materials directly impacts our students' learning experiences. Without access to engaging hands-on activities, cutting-edge technologies, and well-equipped laboratories, it's difficult to spark and sustain their interest in STEM subjects. (Group A)
>
> When students lack the tools and resources to explore STEM concepts, it can lead to frustration, disengagement, and lower achievement. Adequate funding and investment in STEM infrastructure are crucial for supporting student success. (Group C)

The findings from the two groups emphasise the significant role that resources and materials play in enhancing student learning in STEM education. The findings align with Kadziya and Ndebele (2020), who reported a shortage of STEM education resources in Africa, including Zimbabwe. Adequate funding and resource investment are critical to ensuring that all students have access to the necessary tools and infrastructure to succeed, particularly in schools lacking resources. Hands-on activities and technology are essential in making STEM subjects engaging and impactful for students.

Access to high-quality materials, technology, and well-equipped laboratories is essential for engaging students, fostering interest, and ensuring successful outcomes in STEM subjects. Teachers said:

> Providing high-quality, up-to-date lab equipment, technology, and learning materials can greatly impact student engagement and motivation. When students have access to the right tools, they're more likely to be excited about STEM subjects. (Group B)
>
> Ensuring that all schools, regardless of location or socioeconomic status, have access to high-quality STEM facilities, equipment, and instructional materials is essential. Governments and policymakers must prioritise equitable funding and resource allocation to support STEM education in underserved communities. (Group D)

The results from the two groups acknowledge that access to high-quality STEM resources is essential for engaging and motivating students. Group B focuses on the immediate impact of resources on student excitement, while Group D takes a broader view, advocating for equitable resource distribution to ensure that all students, regardless of their background, have equal opportunities for STEM education. Ensuring access to adequate resources in underserved communities is crucial for fostering inclusive and effective STEM learning. The findings are in line with those of Dekeza and Kufakunesu (2017), who advocated for the provision of resources for STEM education.

Community Support and Engagement

Community support is crucial in enhancing student engagement and motivation in STEM education, mainly through local partnerships with parents, businesses, and organisations. In rural areas, the lack of strong community outreach limits students' access to real-world STEM applications and career role models, underscoring the need for active community involvement to bridge this gap and inspire students. Teachers had this to say:

> The level of support and involvement from the local community plays a significant role in shaping our students' STEM engagement and motivation. When we see parents, businesses, and community organisations actively investing in and championing STEM education, it sends a powerful message to our students about the value of these subjects. (Group A)
>
> Unfortunately, we don't have strong community partnerships or outreach programs in rural areas to expose our students to STEM career opportunities and role models. This lack of external support and engagement can limit our students' understanding of STEM skills' relevance and real-world applications. (Group B)

The two groups recognise the importance of community support, but Group B identifies a gap in rural areas that needs addressing. Active community engagement motivates students and connects them to potential careers, making it essential for rural schools to build stronger partnerships to provide students with the exposure and resources they need to succeed in STEM. The involvement of STEM professionals, field trips, and mentorship opportunities plays a vital role in helping students understand the real-world applications of their learning. The findings by Mutseekwa (2021) revealed that during training, teachers were never involved in field trips.

Community engagement and partnerships can bridge the gap between classroom education and career pathways, significantly enhancing student motivation and interest in STEM fields. Teachers said:

> Bringing in STEM professionals, organising field trips, and connecting students with mentors can help them see the real-world applications of what they're learning. These partnerships help students understand the relevance of STEM and get excited about potential careers. (Group C)
>
> Fostering strong community engagement and partnerships will be a key strategy for driving sustained improvements in STEM education within rural secondary schools. By connecting students with STEM professionals, organising field trips, and providing mentorship opportunities, teachers can help students better understand the real-world relevance and potential career pathways in STEM fields. This, in turn, can significantly boost student

motivation, engagement, and interest in pursuing STEM-related studies and careers. (Group D)

Beyond the classroom, we need to expand opportunities for students to engage with STEM-related activities and professionals in their local communities. Partnerships with businesses, universities, and community organisations can provide students with hands-on learning experiences, mentorship, and exposure to STEM career pathways. (Group E)

All three groups agree that fostering community involvement through partnerships, mentorships, and real-world experiences is essential for enhancing STEM education. These approaches help bridge the gap between theoretical knowledge and practical application, motivating students to pursue STEM careers by showing them the relevance of what they are learning. The findings are in line with Kumba et al. (2024), who reported the need for partnerships with various organisations, including industry, to help students engage in real, practical activities where they put theory into practice.

Gender Issues in STEM Education

Gender biases and societal perceptions can significantly impact students' attitudes toward STEM education, often discouraging girls from pursuing these fields. Teachers had this to say:

Cultural and societal perceptions of STEM subjects can shape students' attitudes and aspirations. In most schools, there is still a prevailing notion that STEM fields are more suited for boys, which can inadvertently discourage our girls from pursuing these subjects with the same enthusiasm. (Group C)

Teachers point out a barrier in STEM education regarding gender-based stereotypes that associate STEM subjects with boys. Such perceptions limit girls' aspirations and can lead to a lack of engagement or interest in STEM. The findings align with Maisiri (2024), who reported the need for women to be more represented in STEM fields and prepared for the future of work. It's crucial to challenge these stereotypes and create inclusive, supportive environments where all students, regardless of gender, feel empowered to engage and excel in STEM. Teachers said:

We must actively challenge these outdated stereotypes and promote the message that STEM is for everyone, regardless of gender. Cultivating a more inclusive and supportive STEM culture in our society is crucial for empowering all students to engage and succeed in these vital fields. (Group C)

It's important to actively address stereotypes and biases and create inclusive learning environments where all students, regardless of gender, feel welcomed and supported in STEM. Representation matters, so having diverse role models and mentors can make a big difference. (Group D)

The two groups highlight the importance of challenging stereotypes and creating inclusive, supportive learning environments where all students, regardless of gender, feel encouraged to pursue STEM. This approach emphasises the role of representation through having diverse role models and mentors in STEM who can inspire and motivate students. By fostering a culture of inclusivity, teachers can help break down gender barriers and promote greater diversity in STEM fields. This will benefit students and contribute to a more equitable society where all individuals have equal opportunities to succeed in STEM careers. The findings align with Mabhanda (2016), who highlighted the need for inclusive STEM education practices, aiming to create environments where all students, regardless of socioeconomic status, can engage with and benefit from STEM learning experiences. There is a need to foster diversity and inclusivity within STEM education (Mabhanda, 2016).

Conclusion

In conclusion, STEM education is vital in shaping students' futures by fostering critical thinking, problem-solving, and innovation. However, its successful implementation is influenced by teacher preparedness, resource availability, community support, and addressing gender disparities. The urban-rural divide in STEM education highlights the need for equitable access to resources and opportunities to ensure that all students, regardless of their location, can thrive in STEM fields. Strengthening community involvement, investing in teacher training, and challenging societal stereotypes will be key to creating a more inclusive and effective STEM education system. Addressing these challenges collaboratively would empower the next generation of students with the skills and knowledge necessary to drive progress and innovation in their communities and beyond.

Appendix: Focus Interview Questions

How do teachers conceptualise STEM education in their teaching?/How do teachers define STEM education in their own words?

What factors influence teachers' understanding of STEM as an integrated or subject-specific approach?

What do teachers perceive as the primary benefits of STEM education for students?/ How does STEM education contribute to developing twenty-first-century skills such as problem-solving, critical thinking, and collaboration?

What level of preparedness do teachers feel they have in implementing STEM education?

What types of training or professional development do teachers believe would improve their ability to teach STEM effectively?

How does STEM resources' availability (or lack thereof) impact teaching and learning?

What specific resources do teachers believe would enhance STEM education in their schools?/What challenges do rural teachers face in implementing STEM education compared to their urban counterparts?

What strategies can be implemented to bridge the rural-urban gap in STEM education access and quality?
How does community involvement (parents, businesses, organisations) influence students' participation in STEM education?
What strategies can be employed to strengthen community partnerships and support STEM learning?
How do societal and cultural perceptions influence gender disparities in STEM education?
What measures can encourage more female students to participate in STEM subjects?

References

Adams, K. A., & Lawrence, E. K. (2015). *Research methods, statistics and applications*. Sage.
Baxter, P., & Jack, S. (2010). Qualitative case study methodology: Study design and implementation for novice researchers. *The Qualitative Report, 13*(4), 544–559.
Christensen, L. B., Johnson, R. B., & Turner, L. A. (2015). *Research methods, design, and analysis* (12th ed.). Pearson Education.
Cohen, L., Manion, L., & Morrison, K. (2015). *Research methods in education* (7th ed.). Routledge.
Creswell, J. W. (2015). *A concise introduction to mixed methods research*. Sage.
Curtis, B., & Curtis, C. (2011). *Social research: A practical introduction*. Sage.
Dekeza, C., & Kufakunesu, M. (2017). Implementation of STEM curriculum in rural secondary schools in Zimbabwe: Limits and possibilities. *Journal of Emerging Trends in Educational Research and Policy Studies, 8*(1), 11–15.
Gadzirayi, C. T., et al. (2016). *Diagnostic study on status of STEM in Zimbabwe*. Bindura University of Science Education & Higherlife Foundation. Retrieved December 2024, from https://www.researchgate.net/publication
Gray, D. E. (2011). *Doing research in the real world* (2nd ed.). Sage.
Hennink, M. M. (2014). *Focus group discussions. Understanding qualitative research*. Oxford University Press.
Kadziya, L., & Ndebele, C. (2020). An analysis of the challenges faced by teachers in the delivery of Science, Technology, Engineering and Mathematics education in rural day secondary schools in Zimbabwe. *International Academic Research Journal of Engineering and Technology, 1*(3), 85–95.
Kumba, H., Olanrewaju, O., Akpan, J., Mafuratidze, F., & Ndiyamba, D. (2024). *Engineering education and industry synergy in Zimbabwe: A divide and redirection*. IntechOpen. https://doi.org/10.5772/intechopen.1004886
Mabhanda, W. (2016). Opportunities and factors affecting adoption of STEM education: The case of Gweru Polytechnic first year commerce students. *International Journal of Business Marketing and Management, 1*(5), 1–8.
Madusise, S. (2020). Possibilities for implementation of STEM education in Zimbabwean under-resourced mathematics classrooms. *Journal of African Interdisciplinary Studies, 3*(7), 110–124.
Maisiri, J. (2024). Preparing Zimbabwean women for the future of work: The role of STEM education in the so-called fourth industrial revolution. *South African Review of Sociology, 54*(1), 5–22. https://doi.org/10.1080/21528586.2024.2311399
Mandiudza, M. (2021). Factors motivating female students from Masvingo, Zimbabwe—The choice of taking natural sciences at university. *European Journal of Social Sciences Studies, 6*(2). https://doi.org/10.46827/ejsss.v6i2.1001

Manunure, K., & Leung, A. (2024). Integrating inquiry and mathematical modeling when teaching a common topic in lower secondary school: An iSTEM approach. *Frontiers of Education, 9*(1376951), 1–17. https://doi.org/10.3389/feduc.2024.1376951

Morrison, J. (2006). *TIES STEM education monograph series, attributes of STEM education.* Retrieved from http://daytonos.com/pdf/stem.pdf

Moyo, P. V., & Berdut, I. R. (2017). Interrogating the implementation of STEM education in Zimbabwe. *The International Journal of Humanities and Social Studies, 5*(11), 332–337.

Mutseekwa, C. (2021). STEM practices in Science teacher education curriculum: Perspectives from two secondary school teachers' colleges in Zimbabwe. *Journal of Research in Science, Mathematics and Technology Education, 4*, 75–92. https://doi.org/10.31756/jrsmte.422

Mwembe, D., Zulu, I. L., & Mdlongwa, P. (2022). Needs assessment for facilitators in teaching of STEM subjects in under resourced rural schools: Case of Lupane and Hwange Districts in Zimbabwe. *International Journal of Professional Development, Students and Learning, 4*(1), ep2202. https://doi.org/10.30935/ijpdll/11666

Richardson, V. (1997). *Constructivist teacher education: Building a world of new understandings*. Falmer.

Ruparanganda, F. (2019). Challenges faced by Biology Secondary teachers when using instructional methods that promote Science Technology, Engineering and Mathematic (STEM) education in Zimbabwe. *Texila International Journal of Academic Research, 6*(2), 1–6.

Sarantakos, S. (2013). *Social research*. Palgrave McMillan.

Seale, C. (2012). *Researching society and culture*. Sage.

Silverman, D. (2011). *Interpreting qualitative data* (4th ed.). Sage.

Tsvara, P., Chitate, T., & Johan, J. B. (2018). Examining the impact of Science, Technology, Engineering and Mathematics (STEM) in Zimbabwe's high schools. In *Raising the impact of education research in Africa* (pp. 1–25). https://doi.org/10.4102/aosis.2018.BK53.06

Zhou, M. (2017). *STEM implementation in rural secondary schools: A case of Kwekwe Zhombe district*. Undergraduate research project. Midlands State University.

STEM Education as Praxis to End Poverty: A Curriculum Turn

Abstract Globally, the STEM field has the fastest-growing job market, with more STEM-related work positions than STEM-qualified individuals. Graduates can occupy such STEM-related jobs by gaining qualifications in STEM education. Educational institutions, such as universities and schools, are regarded as venues to advance students' performance in STEM education and, consequently, assist in building a qualified and skilled workforce for the future. Typically, students from underdeveloped communities have limited access to STEM education due to poverty. Their chances of occupying STEM-related jobs are also limited. As a result, they occupy low-paying jobs that propagate the cycle of poverty. Hence, there is a need to increase opportunities for STEM education and career preparedness that break the poverty cycle and allow students to be exposed to STEM fields as well as enter well-compensated careers. As the STEM field is growing and advances in technology are being made, some specialised workers' occupations are becoming less relevant and are being driven into poverty. Such scenarios demonstrate the importance of building foundational STEM education skills from early childhood to the tertiary level. The chapter dwells on how STEM education can help alleviate poverty through curriculum transformation. In pursuing social justice and sustainable development, STEM education can be envisioned as a transformative praxis to address and potentially end poverty. A curriculum turn in STEM education involves reimagining and restructuring the educational framework to emphasise acquiring scientific and technical knowledge and applying STEM skills to tackle real-world issues, including poverty.

Introduction

In the dynamic landscape of the twenty-first century, STEM fields have emerged as the driving force behind global innovation and economic growth (Gadzirayi et al., 2016; Chitate, 2016; Ngwenya et al., 2017). Zimbabwe is witnessing an unprecedented surge in the demand for skilled professionals in STEM disciplines (Chitate, 2016; Nkala et al., 2024). Industries and economies are increasingly reliant on the innovative contributions of individuals proficient in STEM (Nkala et al., 2024). The

B. Chirinda et al., *STEM Education in Poverty*,
https://doi.org/10.1007/978-3-031-92490-3_8

global demand transcends geographical boundaries as nations seek to harness the power of STEM to drive advancements in diverse sectors, including healthcare, information technology, and sustainable energy solutions (Foi & Kean, 2022). Despite the burgeoning demand for STEM professionals, a stark imbalance persists. Job openings in STEM-related fields far exceed the number of qualified individuals entering the workforce (Nguyen et al., 2020). This dissonance poses challenges for industries seeking to fill critical positions and underscores a systemic gap in STEM education and training. The recognition of this imbalance serves as a catalyst for exploring innovative strategies to bridge the divide and empower a more extensive pool of individuals with the skills needed for STEM careers.

At the heart of addressing the demand for STEM professionals lies the pivotal role of STEM education. In Zimbabwe, educational institutions, from primary schools to universities, play a central role in shaping the next generation of STEM professionals (Gadzirayi et al., 2016). In Zimbabwe, the curriculum serves as the foundation for equipping students with the knowledge, skills, and mindset necessary to navigate and contribute to the complex and evolving landscape of STEM disciplines. When strategically designed and implemented, STEM education becomes the conduit through which graduates are prepared for lucrative careers that not only fulfill individual aspirations but also contribute to societal progress and economic development (Moyo & Berdut, 2017).

The chapter explores the multifaceted dimensions of STEM education as a praxis for addressing poverty. Recognising that poverty often limits access to quality education, particularly in STEM fields, the chapter explores how a transformative curriculum turn can break down barriers, empower marginalised communities, and create a more equitable future where the benefits of STEM education are accessible to all. The chapter contributes to the ongoing dialogue on leveraging STEM education for social transformation and economic empowerment.

Access Barriers for Underprivileged Communities in STEM Education: Breaking the Chains of Inequity

In Zimbabwe, most schools are located in rural areas and are government-owned (Gory et al., 2021). In pursuing leveraging STEM education as a transformative force to end poverty, it is imperative to confront the stark reality that underprivileged communities face significant barriers in accessing quality STEM education (Gadzirayi et al., 2016). Poverty, a pervasive challenge in many communities, manifests its impact on education in multifaceted ways. In the realm of STEM education in Zimbabwe, financial constraints often translate into limited access to essential resources, cutting-edge technology, and well-equipped laboratories (Mandiudza, 2021; Zhou, 2017). Underprivileged communities find themselves at a disadvantage when providing students with the necessary tools and infrastructure to engage meaningfully in STEM subjects (Mandiudza, 2021).

Financial constraints act as formidable barriers, shaping the educational landscape for students in underdeveloped communities in Zimbabwe. The costs associated with STEM education, including fees, textbooks, laboratory materials, and technological resources, pose significant challenges for families facing economic hardships (Zhou, 2017). The inability to afford extracurricular STEM activities and supplemental learning resources further exacerbates the disparities (Ngwenya et al., 2017). The Government of Zimbabwe implemented various programs and projects to enhance the quality of STEM teaching and learning (Gadzirayi et al., 2016). One such initiative is the Basic Education Assistance Module (BEAM), which helps underprivileged schoolchildren by covering their tuition, levies, and examination fees. BEAM accomplished this with financial support from the World Bank, UNICEF, and other development partners. In addition, through the UNICEF Education Transition Fund, science kits and textbooks were distributed to all schools in Zimbabwe, significantly reducing the textbook-to-pupil ratio to acceptable levels (Ngwenya et al., 2017). Every secondary school in Zimbabwe receives science kits to promote the teaching and learning of science through practical experiments (Ngwenya et al., 2017).

The correlation between limited access to STEM education and the perpetuation of poverty cycles is critical to understanding the broader societal impact. In Zimbabwe, insufficient exposure to STEM subjects restricts students' ability to pursue higher education and enter STEM-related professions (Gadzirayi et al., 2016). As a consequence, individuals from underprivileged communities may find themselves relegated to low-wage occupations, reinforcing the cyclical nature of poverty. This perpetuation of economic disadvantage is not only detrimental to individual aspirations but also hampers the overall socio-economic development of communities.

As we navigate the complex intersection of poverty and STEM education, it becomes evident that dismantling these barriers requires a holistic approach (Moyo & Berdut, 2017). The United Nations (UN) states that STEM education can reduce poverty and inequality in developing countries, including Zimbabwe (Nguyen et al., 2020). As indicated in chapter "Introduction", Zimbabwe's STEM initiative, introduced at the start of 2016, provides free education to all A-level students registered for STEM subjects like Maths, Computer Science, Physics, Chemistry, and Biology (Ngwenya et al., 2017). This initiative is part of the government's effort to promote the adoption of these crucial subjects, which are central to national development plans (Ngwenya et al., 2017). The Zimbabwean government aims to ensure an adequate number of candidates with science-related skills. The subsequent sections will explore innovative strategies and interventions that can serve as a catalyst for change, transforming the landscape of STEM education for underprivileged communities. By addressing the root causes of limited access, we aim to break the chains of inequity and pave the way for a future where STEM education is truly inclusive, accessible, and transformative for all.

Importance of Early Childhood STEM Education: Nurturing Curiosity and Building Foundations

In the overarching endeavour to utilize STEM education as a praxis to end poverty, the significance of early childhood education emerges as a pivotal focal point. In Zimbabwe, researchers such as Gadzirayi et al. (2016) have noted that for sustainability, the interest of students in STEM needs to be harnessed from as early as preschool and nurtured along the way. To successfully promote STEM education, initiatives should begin at the grassroots level by introducing children in primary school, and even those in preschool, to the fundamentals of technology from an early age (Mabhanda, 2016). Developing students' STEM competence at a young age is paramount for nurturing innovation talents (Dube, 2018). Advocating for early STEM education is linked to its substantial impact on preparing students for the challenges of the twenty-first century and their future success in STEM fields. Given that everyone faces life problems, the ability to identify and solve these issues should be taught from primary school onwards (Moyo & Berdut, 2017).

Additionally, the transition between educational levels must be smooth, ensuring seamless alignment from Early Childhood Development to university (Pakombwele & Tsakeni, 2022). Each level of education should build on the previous one, with the primary school curriculum laying a solid foundation for secondary education, preparing students for tertiary education (Moyo & Berdut, 2017; Pakombwele & Tsakeni, 2022). As the STEM centre at Bindura University aptly notes, introducing STEM education at the infant level aims to provide students with a thorough grounding in STEM from an early age, acquainting them with various STEM-related fields and occupations as well as familiarising them with the STEM way of learning (Moyo & Berdut, 2017).

The Government of Zimbabwe implemented various programs and projects to enhance the quality of STEM teaching and learning (Gadzirayi et al., 2016). One significant effort was the curriculum review, which integrated science and mathematics into the formal school curriculum starting from the Early Childhood Development (ECD) level. The formative years of a child's education serve as the building blocks for cognitive development and skill acquisition. Early childhood education, particularly in STEM subjects, is a cornerstone for building foundational skills that lay the groundwork for future academic and professional pursuits (Dube, 2018). Cognitive abilities, problem-solving skills, and spatial awareness -integral components of STEM proficiency - are nurtured during these early years (Mugweni, 2016; Dube, 2018). This examination underscores the long-term impact that early exposure to STEM concepts can have on a child's educational trajectory.

Early childhood is a time of innate curiosity and rapid cognitive development. Introducing STEM concepts during this critical period capitalises on a child's natural inclination to explore and fosters a lasting interest in science, technology, engineering, and mathematics (Mabhanda, 2016; Dube, 2018). It is agreed that STEM subjects should be taught as early as possible, as early experiences and engagement in these fields stimulate concept development in young students (Mugweni, 2016). Early exposure to hands-on STEM activities nurtures a sense of wonder and curiosity, laying the foundation for a positive attitude towards learning. The significance of igniting this curiosity early on is profound, as it instils a lifelong appreciation for STEM disciplines and opens doors to a myriad of educational and career possibilities (Dube, 2018). Hence, in Zimbabwe, there is a need for widespread marketing of STEM education, starting from preschools and continuing up to the highest possible level.

At the junior school level, emphasis is placed on developing STEM disciplines through inquiry-based, project-based, and hands-on learning that integrates vital elements from various subjects in natural and engaging ways (Gondo & Mudekunye, 2020). In Zimbabwe play-based teaching methods in early childhood, an increasing number of studies highlight how early STEM skills are observed and nurtured through play-based learning (Marazanye, 2024). Pakombwele and Tsakeni (2022) assert that children inherently exhibit scientific behaviors during free play, engaging in observation, questioning, and problem-solving, effectively demonstrating STEM principles in action. Early childhood professionals widely recognize that discovery and exploration through play are fundamental to STEM education (Marazanye, 2024). Additionally, the integration of digital technologies into intentional teaching and play-based learning has further advanced early STEM education practices (Pakombwele, 2024).

Curriculum Transformation in STEM Education: Paving the Way for Equitable Access and Social Impact

Within the overarching framework of using STEM education as praxis to combat poverty, a critical component lies in the transformative shift of the curriculum. Traditional STEM curricula have often been criticised for their insularity and focus on theoretical knowledge divorced from real-world applications (Zhou, 2017). A curriculum turn is essential to address this gap and align STEM education with the dynamic needs of contemporary society. There is a need to reevaluate and reshape STEM curricula to be more responsive, inclusive, and purpose-driven. A forward-looking curriculum turn is not merely a pedagogical shift but a strategic move to empower students with the skills needed for professional success and societal impact. In 2016, the Ministry of Primary and Secondary Education (MoPSE) launched a new curriculum for both primary and secondary school levels (MoPSE, 2016). A primary objective of this curriculum is to develop twenty-first-century skills in students, including problem-solving, creativity, innovativeness, and critical thinking (Ministry of Primary and Secondary Education: Zimbabwe Education Blueprint: 2015–2022). These skills align with the objectives of the STEM initiative, which aims to train and develop cutting-edge skills to support Zimbabwe's industrialization and global competitiveness (Moyo, 2019; Zimbabwe Curriculum

Framework (ZCF), 2015–2022). The initiative seeks to cultivate a critical mass of scientific mindsets empowered to create new technologies, industries, jobs, markets, and practical solutions to current and future challenges (Moyo & Berdut, 2017). "STEM subjects are the headlights for the country's future. Countries that have overcome extreme poverty, massive disease burdens, unemployment, political turmoil, and social disorientation have embraced STEM" (Moyo & Berdut, 2017, p 332).

A pivotal facet of reimagining STEM education as a praxis to end poverty involves intentionally designing a community-centric curriculum. This transformative approach entails tailoring the educational experience to meet the specific needs and challenges of local communities, especially those grappling with poverty. An innovative curriculum turn involves a deliberate shift towards real-world problem-solving and community-centric content (Madusise, 2019). By integrating authentic challenges into the curriculum, students are exposed to the complexities of STEM fields and learn to apply their knowledge in meaningful ways. This approach fosters a sense of relevance and purpose, making STEM education more engaging and aligning it with the practical challenges communities face (Madusise, 2019). Whether through project-based learning, case studies, or collaborative initiatives with local communities, these innovative approaches redefine the learning experience.

An integral aspect of the curriculum turn in STEM education involves the strategic integration of real-world problem-solving. This transformative approach shifts the educational paradigm from theoretical abstraction to practical application (Parawira, 2016; Madusise, 2019), emphasising the direct relevance of STEM skills in addressing tangible issues, particularly those related to poverty. There is a need to design a curriculum aligned with local needs. The community-centric curriculum is not a one-size-fits-all model; rather, it is a dynamic and responsive framework designed to address the unique needs of each local community. Educators collaborate with stakeholders to identify and understand the challenges residents face, particularly those related to poverty. This collaborative process ensures that the curriculum is relevant, contextual, and aligned with the specific socio-economic dynamics of the community (Madusise, 2019).

At the heart of this curriculum transformation lies the incorporation of community-centric content. The curriculum is enriched with case studies, examples, and projects directly relating to the local context (Madusise, 2019; Mashoko, 2022). Whether it involves studying the impact of environmental factors on agriculture, developing technologies for affordable healthcare, or addressing infrastructure challenges, the curriculum becomes a tool for immersing students in the real-world issues faced by their communities (Alkair et al., 2023). By delving into community-centric content, students develop a profound understanding of the societal implications of STEM (Chirinda et al., 2023). They learn to contextualise theoretical knowledge within the fabric of their community, recognising the potential of STEM solutions to effect positive change (Chirinda et al., 2023). This deeper understanding goes beyond academic achievement, fostering a sense of responsibility and of STEM's role in shaping the well-being of individuals and communities (Dube, 2018; Madusise, 2019).

The community-centric curriculum positions students as active contributors to sustainable solutions. In Zimbabwe, rather than viewing STEM as a detached academic pursuit, students are encouraged to see themselves as agents of change within their communities (Chitate, 2016; Moyo & Berdut, 2017). Through hands-on projects and collaborative initiatives, they apply their STEM skills to devise practical solutions that directly impact poverty reduction (Chitate, 2016). This experiential learning approach instils a sense of agency, empowering students to believe in their capacity to make a meaningful difference. Beyond technical proficiency, a community-centric curriculum nurtures civic engagement and community empowerment. Students recognise their roles as responsible citizens actively participating in the betterment of their communities (Moyo & Berdut, 2017). This ethos of civic responsibility becomes a guiding principle, fostering a lifelong commitment to addressing societal challenges through applying STEM skills (Moyo & Berdut, 2017).

The traditional dichotomy between classroom learning and real-world application is blurred as STEM education embraces a paradigm shift (Parawira, 2016). The curriculum is enriched by the infusion of real-world problems, creating a dynamic learning environment where students actively engage with their communities' challenges (Parawira, 2016). Such integration ensures that STEM education transcends theoretical boundaries, preparing students to become proactive contributors to societal progress. Real-world context-based teaching and learning aim to connect classroom learning with students' real-life experiences and situations (Madusise, 2019). The approach recognises that students are more engaged and motivated when they can see the relevance of what they are learning to their daily lives. By integrating real-world contexts into the curriculum, educators can create meaningful learning experiences that help students understand the practical applications of STEM concepts (Madusise, 2019).

A key focus of the curriculum turn is to emphasise projects and activities that directly tackle poverty-related issues in Zimbabwe (Ngwenya et al., 2017). By centring STEM education on addressing real-world challenges, particularly those linked to socio-economic disparities, students gain a deeper understanding of the complexities surrounding poverty. Projects may range from developing innovative solutions for sustainable agriculture and clean energy to creating technology-driven initiatives that enhance access to education and healthcare in underserved communities. Beyond theoretical comprehension, the curriculum encourages students to actively apply STEM concepts to devise solutions for the identified problems in Zimbabwe (Mpofu, 2019). The hands-on application fosters a deeper mastery of STEM skills and promotes critical thinking, creativity, and collaboration (Parawira, 2016). Students are exposed to the intricacies of problem-solving and equipped with the tools to enact positive change in their communities (Madusise, 2019).

The curriculum transformation bridges the gap between theoretical knowledge and practical application in Zimbabwe (Kumba et al., 2024; Mutseekwa, 2021). Students see the direct relevance of their STEM education to the challenges faced by their communities, making the learning experience more meaningful and motivating (Madusise, 2019). This connection between theory and practice enhances

STEM education's academic and practical dimensions. The practical relevance of STEM skills is accentuated through real-world problem-solving (Parawira, 2016). Students witness firsthand how their knowledge can be harnessed to create meaningful and sustainable solutions. This emphasis enhances their confidence in applying STEM concepts and reinforces that STEM skills are potent tools for effecting positive change in the world. The integration of real-world problem-solving extends beyond the classroom, cultivating a culture of lifelong learning and civic engagement (Madusise, 2019). Students with a sense of purpose are more likely to continue exploring ways their STEM skills can contribute to social betterment beyond their formal education.

Real-world problem-solving within the context of poverty inherently instils a sense of social responsibility among students (Mpofu, 2019). By engaging with issues that directly impact individuals facing economic challenges, students develop empathy and an understanding of their role as STEM practitioners in addressing societal inequalities (Chitate, 2016). This ethos of social responsibility becomes a guiding principle, influencing students' perspectives and choices throughout their educational and professional journeys.

A community-centric curriculum is a powerful tool in the curriculum turn, elevating STEM education from a detached academic pursuit to a dynamic force for societal transformation. By immersing students in the specific needs of their communities and empowering them to contribute actively to poverty reduction, this approach not only enriches their educational experience but also positions STEM as a driving force in the pursuit of social equity and community well-being. Integrating real-world problem-solving into the STEM curriculum marks a pivotal stride in the curriculum turn. Educators foster a generation of socially responsible and adept problem-solvers by empowering students to apply STEM concepts to address poverty-related challenges. This approach not only transforms the educational experience but also positions STEM education as a potent praxis to actively contribute to the alleviation of poverty and the pursuit of a more equitable world.

The Role of Interdisciplinary Connections in STEM Curriculum Design

The traditional approach to education often compartmentalises disciplines, isolating STEM subjects from other areas of knowledge (Zhou, 2017; Moyo, 2019). This approach compartmentalises these subjects instead of promoting their integration, hindering the development of educators who can effectively teach interdisciplinary concepts (Manokore & Sibanda, 2024). The lack of integration of STEM in institutions in various countries perpetuates a silo instructional approach (Mpofu, 2019). Consequently, this instructional method fails to foster situated learning, which occurs in real-world contexts, thereby inhibiting students' interest and engagement in STEM fields. Zhou (2017) emphasises the persistence of a silo instructional approach, where science and mathematics are often taught separately without considering their interdisciplinary nature. This method fails to capitalise on the potential synergies between these subjects and neglects collaborative learning and problem-solving opportunities.

Breaking down the silos between STEM subjects, interdisciplinary connections are pivotal in offering a holistic understanding of complex issues (Zhou, 2017; Makonye & Moodley, 2023). In the quest to redefine STEM education as a praxis for poverty alleviation, one key element of the curriculum turn is the strategic incorporation of interdisciplinary connections. This transformative approach seeks to break down the silos between STEM subjects, recognising the interconnectedness of knowledge and fostering a holistic understanding of complex societal issues, as reported by Makonye and Moodley (2023) in South Africa. Interdisciplinary connections shatter these silos, creating a fluid and interconnected learning environment (Makonye & Moodley, 2023). There has been a shift from separate subject specialism towards an integrated science curriculum in Zimbabwe, reflecting future trends. The new Zimbabwean course emphasises practical applications of science and technology currently in use in the country, contrasting with conventional approaches that first study scientific concepts and principles before relating them to real-world applications. The course begins with applications of science in agriculture, industry, energy uses, structures and mechanical systems, and the community. These areas provide the focus for all secondary school core science, extended science and combined science up to the Ordinary level.

In Zimbabwe, practical learning areas such as design and technology, information and communication technology (ICT), art, and theatre arts are included to broaden the Zimbabwean educational base at all levels (Ministry of Primary and Secondary Education Curriculum Review Process Narrative Report, 2014–2015; Gondo & Mudekunye, 2020). Through STEM, students investigate and attempt to answer various questions, fostering their curiosity and developing a strong foundation in STEM inquiry (Gondo & Mudekunye, 2020; Makonye & Moodley, 2023). The curriculum design intentionally blurs the boundaries between science, technology, engineering, and mathematics, fostering a more integrated and nuanced understanding of real-world challenges, including those related to poverty. Incorporating elements from social sciences, humanities, and the arts into STEM education enriches students' perspectives and prepares them to address challenges with a comprehensive toolkit (Gondo & Mudekunye, 2020).

Bybee (2010) highlights the importance of integrating STEM disciplines to provide students with a holistic understanding of how these subjects intersect and apply in practical scenarios. However, the current approach isolates each subject, limiting students' ability to grasp their interconnectedness and applicability in solving real-world problems (Moyo, 2019; Mpofu, 2019). In Zimbabwe, STEM is rapidly becoming a core educational initiative, spanning from Early Childhood Education to the tertiary level, due to its integrative approach (Gondo & Mudekunye, 2020; Pakombwele & Tsakeni, 2022). The country's contribution to the new global economy relies on an education system with a STEM focus, aiming to effectively empower citizens for the twenty-first century (Manokore & Sibanda, 2024).

Regarding STEM education, complex societal issues, such as poverty, demand multifaceted solutions that extend beyond the scope of any single discipline (Mpofu, 2019). Interdisciplinary connections give students a broader perspective, allowing them to view challenges through various lenses (Mpofu, 2019). In Zimbabwe, by incorporating elements from social sciences, humanities, and the arts into STEM education, students gain a more comprehensive understanding of the interconnected nature of societal issues, recognising the intricate web of economic, cultural, and environmental factors at play (Gadzirayi et al., 2016).

Better Environment Science Teaching (BEST) is the cornerstone of science teaching at the primary level in Zimbabwe (Risiro, 2014). This program led to the development of a new primary science syllabus and the in-service training of teachers, heads, District Education Officers, and Education Officers (Gudyanga et al., 2013). One of its key advantages is using the environment as a laboratory for biological and physical sciences (Risiro, 2014). Zimbabwe leads the conceptual framework and implementation of the project, while the German Agency for Technical Cooperation provides funding.

In Zimbabwe, cross-cutting concepts are crucial in linking STEM disciplines and providing a framework for understanding the interconnectedness of scientific phenomena (Mpofu & Mutseekwa, 2023). Two such cross-cutting concepts highlighted by Madusise (2019) are creative design and the cultural embeddedness of STEM. Creative design involves the application of scientific principles and knowledge to develop innovative solutions to real-world problems. It emphasises the iterative process of problem-solving, where scientific discoveries serve as a foundation for designing new technologies, products, or systems. Creative design bridges the gap between science and engineering, fostering creativity, critical thinking, and collaboration among students (Madusise, 2019; Mutseekwa, 2021). By incorporating these cross-cutting concepts into STEM education, educators can provide students with a more holistic understanding of how scientific concepts, engineering principles, and societal factors intersect and influence one another. This interdisciplinary approach encourages students to think critically about the complex relationships between science, technology, society, and culture, empowering them to become informed and responsible global citizens (Mutseekwa, 2021).

In Zimbabwe, the cultural embeddedness of STEM recognises that STEM knowledge and practices are shaped by cultural and societal contexts (Chitate, 2016). Humans construct scientific knowledge through interactions with the natural world and develop technologies that reflect their cultural values, beliefs, and priorities (Madusise, 2019; Matindike & Ramdhany, 2024). Understanding the cultural embeddedness of STEM disciplines is essential for promoting diversity, equity, and inclusion in STEM education and ensuring that scientific advancements benefit all members of society (Matindike & Ramdhany, 2024). Problem-solving within the STEM curriculum is enriched by considering the cultural context. Students learn to tailor solutions to diverse cultural settings, acknowledging that the effectiveness of STEM interventions can vary based on cultural nuances (Madusise, 2019).

Exposure to interdisciplinary connections enriches students' perspectives, challenging them to think beyond the narrow confines of their primary STEM focus.

Integrating insights from diverse fields broadens their worldview, fostering critical thinking and adaptability (Makonye & Moodley, 2023). Students are encouraged to explore how their STEM knowledge intersects with and complements insights from other disciplines, creating a richer intellectual tapestry. Interdisciplinary connections aim to broaden students' intellectual horizons and equip them with a comprehensive toolkit for addressing real-world challenges (Mpofu & Mutseekwa, 2023). As a complex and systemic issue, poverty requires solutions that consider economic, social, and environmental dimensions. By incorporating elements from disciplines beyond STEM, students are better prepared to approach challenges with a holistic understanding and develop innovative, sustainable solutions (Mpofu & Mutseekwa, 2023). Interdisciplinary connections naturally integrate ethical and social considerations into STEM education (Madusise, 2019). As students navigate the intersections of various disciplines, they are prompted to consider the ethical implications of their work and the social impact of STEM solutions. This ethical framework becomes integral to their decision-making process, emphasising the responsible application of STEM knowledge for the greater good.

The collaborative nature of interdisciplinary connections encourages students to engage in collaborative problem-solving (Makonye & Moodley, 2023). Working with peers from different STEM backgrounds and incorporating insights from non-STEM disciplines fosters teamwork and collaboration. This collaborative ethos reflects the real-world dynamics of addressing societal challenges, where diverse teams bring varied perspectives to create innovative solutions.

Interdisciplinary connections in curriculum design are paramount in transforming STEM education into a praxis for addressing poverty. By breaking down disciplinary silos, enriching perspectives, and preparing students with a comprehensive toolkit, interdisciplinary connections empower the next generation of STEM practitioners to contribute meaningfully to the complex challenges of our interconnected world. This approach enhances the educational experience and positions STEM as a dynamic force capable of creating positive societal change.

The Role of Ethical Considerations in STEM Curriculum Design

As STEM education undergoes a transformative curriculum turn, integrating interdisciplinary connections becomes a matter of enriching academic perspectives and an essential avenue for addressing ethical considerations. As technological advancements raise ethical concerns, the curriculum must incorporate discussions on responsible innovation, environmental sustainability, and the ethical use of technology. Integrating ethical considerations into STEM education ensures that students develop a consciousness of the societal impact of their work.

The rapid pace of technological advancements brings forth ethical concerns that demand thoughtful consideration. A curriculum turn recognises the imperative to address these concerns head-on. Rather than viewing technology in isolation, students are encouraged to critically examine the ethical implications of innovations. This includes considerations of privacy, data security, and the potential societal impacts of emerging technologies. Responsible innovation becomes a cornerstone of the curriculum. Students engage in discussions beyond the technical aspects of their projects, delving into questions about how their innovations may impact communities and individuals. This forward-thinking approach instils a sense of responsibility from the inception of an idea to its implementation, emphasising that ethical principles and considerations should guide innovation.

Environmental sustainability is woven into the fabric of the curriculum, recognising the role of STEM in shaping a sustainable future. Discussions and projects explore how STEM solutions can contribute to environmental conservation, renewable energy, and climate change mitigation (Alkair et al., 2023). Students are challenged to think critically about the ecological footprint of their work and explore ways in which STEM can be a force for positive environmental change (Nguyen et al., 2020). In a world increasingly reliant on technology, the ethical use of technology becomes a central theme in the curriculum. Students are guided to consider the potential ethical dilemmas associated with technological applications, from artificial intelligence to biotechnology. Emphasising responsible and ethical practices in the development and deployment of technology ensures that STEM practitioners are mindful of the broader societal implications of their creations (Karimi & Pina, 2021).

A curriculum turn positions the ethical framework as a foundational element for decision-making in STEM. Students are equipped with technical skills and a moral compass that guides them in navigating the complex ethical terrain of STEM fields. This emphasis on ethical decision-making prepares students to be responsible contributors to society, capable of balancing innovation with a profound understanding of its ethical implications. Incorporating ethical considerations into the STEM curriculum within the context of interdisciplinary connections is pivotal in shaping responsible and conscientious STEM practitioners. By addressing ethical concerns associated with technological advancements, promoting environmental sustainability, and emphasising the ethical use of technology, a curriculum turn ensures that STEM education becomes a praxis for positive societal impact. This transformative approach enriches the educational experience and positions STEM as a force for ethical innovation and responsible advancement.

The Role of Social Entrepreneurship Education in STEM Curriculum Design

According to a study in Zimbabwe by Nkala et al. (2024), empowering students with a mindset for social entrepreneurship is integral to a curriculum turn. Entrepreneurship is pivotal in driving innovation, economic growth, and societal advancement, particularly within STEM fields (Nkala et al., 2024). Regarding the Zimbabwean New Curriculum Framework, Mapfumo (2015) highlighted several emerging issues from the consultation process, such as the promotion of enterprise development and the emphasis on teaching sciences, mathematics, technology, technical/vocational subjects, and ICT. Zimbabwe has introduced a new curriculum in its primary and secondary school systems based on recommendations from the Presidential Commission Report on Education and Training of 1999 (Saiden, 2017). The commission found that the education system did not emphasise self-reliance and entrepreneurship, nor did it strongly promote teaching Science, Mathematics, Technology, Vocational and Technical subjects, and local languages (Chitate, 2016). It recommended prioritising STEM and entrepreneurship (Nkala et al., 2024). The new education curriculum aims to equip today's generation with the competencies, knowledge, skills, and attitudes necessary for driving the country's socio-economic growth. The focus is now on creating entrepreneurs who can generate employment, thus addressing the issue of unemployed graduates (Madusise, 2019). This means incorporating elements of entrepreneurship into the new curriculum. The New Curriculum was crafted with the understanding that education is fundamental to personal and national development. It aimed to develop a highly skilled and innovative workforce essential for social, cultural, and economic growth (Saiden, 2017). Studies by Saiden (2017) and Nkala et al. (2024) in Zimbabwe revealed that while the new curriculum includes entrepreneurship skills, these need further enhancement.

By emphasising the role of STEM in Zimbabwe in creating solutions for societal challenges, students learn the technical aspects and the ethical and social dimensions of their work (Mabhanda, 2016). Social entrepreneurship education instils a sense of responsibility and commitment to positively impacting the world. Social entrepreneurship education begins by instilling in students a mindset beyond traditional STEM education (Mabhanda, 2016; Masunda et al., 2018). It empowers them to view their technical skills as tools for creating positive societal change. This shift in perspective encourages students to explore ways their STEM knowledge can be harnessed to address pressing social issues, including those related to poverty (Saiden, 2017).

Entrepreneurship courses have been introduced in Zimbabwean STEM programs at the university level (Masunda et al., 2018; Nkala et al., 2024).). A fundamental tenet of social entrepreneurship education is the emphasis on the pivotal role of STEM in creating solutions for societal challenges (Mabhanda, 2016; Saiden, 2017). Students are guided to recognise the transformative potential of their STEM skills in developing innovative solutions that contribute to social betterment (Masunda et al., 2018). According to Nani (2016), school administrators and parents need to be made aware of the importance of instilling the right attitudes in students regarding the value of entrepreneurship. This recognition becomes a driving force in shaping their educational and professional goals.

In Zimbabwe, entrepreneurship education aligns STEM education with a broader purpose—positively impacting the world (Saiden, 2017; Masunda et al., 2018). Students are encouraged to view their STEM careers as avenues for personal success and opportunities to effect positive change (Saiden, 2017). This commitment to societal impact becomes a driving force, motivating students to actively seek ways in which their skills can contribute to poverty reduction and other social challenges. A curriculum turn integrates social entrepreneurship education with broader social and environmental goals (Saiden, 2017). By aligning STEM education with these overarching objectives, students are exposed to a vision where their skills can be leveraged to create solutions that address poverty, inequality, and environmental sustainability (Saiden, 2017). This alignment reinforces the interconnectedness of STEM disciplines with broader societal needs.

The Business Education Partnership of Zimbabwe (BEPAZ) is a collaborative initiative between the Ministry of Education and the business community aimed at promoting technical and vocational education within the school system (Gudyanga et al., 2013). Through this partnership, businesses contribute financial resources to enhance teaching and training in fields with a technical and vocational focus. This support is channelled into programs such as "School on the Shop Floor" and "Teacher Placement," which are designed to expose both secondary school students and their teachers to the working world in general and to new technologies in particular (Gudyanga et al., 2013).

As part of interdisciplinary connections, entrepreneurship education elevates STEM education to a realm where technical proficiency is seamlessly integrated with ethical and social consciousness (Nkala et al., 2024). This transformative approach enriches students' educational experiences and positions STEM as a dynamic force that fosters positive societal change. Through this integrated lens, the next generation of STEM practitioners emerges not only as experts in their fields but as socially responsible innovators committed to addressing the complex challenges of our world. In Zimbabwe, developing entrepreneurship skills at the early stages of learning significantly contributes to a country's economic development. It lays a solid foundation for creativity, nurturing, and innovation of business ideas that can be used later in life (Saiden, 2017). Introducing entrepreneurship at the tertiary level may be too late, as basic entrepreneurial concepts are better taught at a young age when students are in their discovery, manipulative, and experimental stages. Conceptualising a business idea should start early and be nurtured until it develops into a sustainable venture (Nani, 2016). This process can begin in primary school and continue through high school. By the time students graduate from high school, they will have matured into budding entrepreneurs. If they pursue entrepreneurship at the university level, they will be well-prepared to enter the world as fully-fledged entrepreneurs capable of establishing successful business entities (Nani, 2016).

Curriculum Transformation in STEM Education: Inclusivity and Diversity

Fostering an inclusive and diverse learning environment within the STEM curriculum is important as it addresses gender, socioeconomic, and racial disparities by highlighting the contributions of individuals from diverse backgrounds to STEM fields. Creating a curriculum that celebrates diversity promotes an inclusive STEM culture, ensuring that all students feel welcome and represented (Palid et al., 2023). As STEM education undergoes a curriculum turn, a pivotal aspect is the intentional integration of inclusivity and diversity (Dekeza & Kufakunesu, 2017). This transformative approach recognizes the significant impact of inclusive curriculum design on breaking gender, socioeconomic, and racial disparities in STEM fields (Palid et al., 2023).

The government of Zimbabwe aims to achieve equal participation of men and women in STEM programs across all levels of education (Gweshe & Chiware, 2023; Garwe & Chikwiri, 2021). Recognising the critical role of higher education in developing discipline-specific, professional, and leadership skills and being aware of the need to advance government efforts in closing the gender gap, the Women's University in Africa (WUA) (Garwe & Chikwiri, 2021). The University began operating as a private institution in 2002, primarily focused on providing mature women with a second chance to acquire higher education qualifications, empowering them to fulfil socio-economic, political, and leadership roles (Garwe & Chikwiri, 2021). According to the WUA founders, the empowerment of women is crucial for achieving sustainable economic growth (Nherera, 2015). WUA's Charter allows it to recruit students and staff in a ratio of 17 women to three men (Garwe & Chikwiri, 2021).

Inclusive curriculum design is a powerful tool for breaking gender, socioeconomic, and racial disparities in STEM education (Chingarande et al., 2023; Dekeza & Kufakunesu, 2017). By reimagining the curriculum to be accessible, representative, and culturally sensitive, educators create an environment that fosters equal opportunities for all students. This intentional design addresses historical imbalances and provides a foundation for a more equitable STEM education landscape. Representation matters, especially in STEM fields where certain groups may have been historically underrepresented. A curriculum turn emphasises the inclusion of diverse perspectives, experiences, and achievements (Dekeza & Kufakunesu, 2017). Highlighting diverse role models from various genders, socioeconomic backgrounds, and races serves as inspiration for students. Seeing individuals who look like them and have succeeded in STEM fields encourages underrepresented groups to envision themselves as future STEM professionals.

Creating an inclusive learning environment within STEM education involves implementing strategic and intentional practices that foster a sense of belonging and support for all students. The curriculum actively promotes a culture of inclusivity that values and respects diverse perspectives. Students are encouraged to share their experiences, fostering a sense of belonging and acceptance. Inclusive curriculum design involves adapting teaching materials to reflect diverse voices, cultures, and historical contributions (Madusise, 2019). This ensures that students from various backgrounds see themselves represented in the curriculum. Culturally responsive teaching acknowledges and leverages students' cultural backgrounds to enhance the learning experience (Madusise, 2019). Integrating students' cultural references and experiences into the curriculum makes it more relatable and engaging (Madusise, 2019).

Ensuring that learning resources are accessible to all students is crucial for inclusivity in Zimbabwe (Dube et al., 2021). This includes considering factors such as language proficiency, disabilities, and socio-economic constraints (Dube et al., 2021). Inclusive learning environments promote collaboration and teamwork among students with diverse backgrounds. Group projects and activities encourage students to appreciate and learn from each other's unique perspectives. A curriculum turn includes targeted support for underrepresented groups (Dekeza & Kufakunesu, 2017). This may involve mentorship programs, networking opportunities, and initiatives that address the challenges these groups face in adaptation to STEM education (Kadziya & Ndebele, 2020). An inclusive and diverse STEM curriculum is not static; it is a dynamic entity that requires continuous evaluation and adaptation (Dekeza & Kufakunesu, 2017; Moyo, 2019). Regular assessment ensures that the curriculum remains responsive to the evolving needs of a diverse student body. Feedback mechanisms, surveys, and ongoing dialogue with students and educators contribute to improving the curriculum's inclusivity and diversity.

Integrating inclusivity and diversity within the STEM curriculum is integral to a curriculum turn, fostering an environment where all students, regardless of gender, socioeconomic status, or race, can thrive. By breaking traditional disparities, encouraging representation, and implementing inclusive strategies, STEM education becomes praxis for equity and empowers a new generation of diverse and capable STEM professionals.

Curriculum Transformation in STEM Education: Digital Literacy and Access

Within the framework of reimagining STEM education as a praxis to end poverty, the integration of digital literacy and equitable access to technology stands out as a crucial aspect of the curriculum turn. Digital literacy is foundational to modern STEM education, equipping students with the skills to navigate, critically assess, and utilise digital technologies (Baterna et al., 2020). A curriculum turn acknowledges the evolving nature of the workforce, where proficiency in digital tools is not only advantageous but often essential. Digital literacy encompasses a range of skills, including information literacy, computational thinking, and proficiency in using digital collaboration tools, all of which are integrated into the STEM curriculum (Hrynevych et al., 2022).

The digital divide, characterised by disparities in access to technology and digital resources, poses a significant challenge to inclusive STEM education in Zimbabwe (Chiremba, 2023). A widening technological divide between rural and urban schools in Zimbabwe is evident, with urban schools advancing in line with the Fourth Industrial Revolution, while rural schools are still lagging behind, struggling to catch up with the Second Industrial Revolution (Chiremba, 2023; Moyo, 2022). A curriculum turn recognises the importance of addressing this divide to prevent exacerbating existing inequalities. Mobile apps can bridge the rural-urban divide in deprived learning situations in Zimbabwe (Maphosa et al., 2021). This approach is viable, as primary and secondary education in Zimbabwe has increasingly integrated ICT alongside STEM education. The feasibility of these approaches is enhanced by the country's dedicated national ICT policy, adopted in 2005, which significantly promotes the use of ICT in education, including its pedagogical applications in educational institutions (Isaacs, 2007). Therefore, it is crucial to fully operationalize this policy to maximize the benefits of ICT in all learning institutions.

Additionally, various initiatives can be integrated into the curriculum to bridge the digital divide and empower all students with digital skills. Designing a curriculum that incorporates digital literacy as a core component, provides students with hands-on experiences using digital tools, and fosters a culture of digital exploration is important for Zimbabwe (Mabhanda, 2016). There is a need to implement programs that introduce and integrate technology into various subjects, allowing students to apply digital skills in real-world contexts (Mabhanda, 2016). This goes beyond standalone digital literacy courses and embeds technology seamlessly into the learning experience. Also, collaboration with private organisations to provide resources, such as donated or subsidised devices, internet connectivity solutions, and mentorship programs, to schools and students in need is useful (Mabhanda, 2016).

In Zimbabwe, efforts from the private sector and government are being made to harness the power of technology in teaching and learning at all educational levels (Chirume & Kwabena, 2016). The Ministry of Primary and Secondary Education (MoPSE) has also partnered with Econet Ruzivo Digital Learning and Wagona Maths, a mathematics tutorial platform, to enhance ICT integration in the learning process and an online interactive platform providing e-learning services in Zimbabwe (Zambaza, 2022). This platform is available free of charge to those who can access it via the internet and is designed for primary and secondary school students. Additionally, Econet Wireless has facilitated free access to the Ruzivo platform when using Econet cell phone lines (UNICEF, 2020). However, these platforms require a computer or smartphone and internet connectivity to access the learning material. While introducing these learning platforms is a commendable effort to complement ICT integration, with less than 25% of schools having internet connectivity, it remains a distant reality for many (Chiremba, 2023). During the COVID-19 pandemic, the education partner UNICEF produced 40 TV and 1000 radio lessons, targeting 1.7 million students (Chiremba, 2023). Disadvantaged students are being systematically excluded from participating in ICT-integrated learning experiences. Without access to a radio, television, or smartphone, which is the case for most rural homes in Zimbabwe, students cannot access online lessons, even though they have been made available (Chiremba, 2023).

Integrating digital literacy and ensuring equitable access to technology within the STEM curriculum are integral components of the curriculum turn. By addressing the digital divide and empowering all students with essential digital skills, STEM education becomes a praxis for preparing students to thrive in the digital age, breaking down barriers to access, and fostering a more inclusive and equitable educational landscape.

Curriculum Transformation in STEM Education: Experiential Learning Opportunities

Integrating experiential learning opportunities plays a pivotal role in a curriculum turn towards utilising STEM education as a praxis to end poverty. In Zimbabwe, internships offer students the opportunity to apply theoretical knowledge gained in the classroom to real-world scenarios (Nani, 2016). In the context of poverty alleviation, STEM students can intern with organisations, non-profits, or government agencies working on projects related to sustainable development, healthcare, or education (Nkala et al., 2024; Kalolo, 2016). Internships with organisations focused on renewable energy can provide hands-on experience in developing sustainable energy solutions for poverty-stricken areas.

Apprenticeships involve a more immersive and extended learning experience, allowing students to work closely with professionals in their chosen field (Nkala et al., 2024). Apprentices working with engineering firms may contribute to designing and implementing infrastructure projects to improve living conditions in impoverished regions (Attard et al., 2021). In Zimbabwe, students interning with technology companies can be involved in developing software or applications that address challenges related to education, access to information, or economic opportunities (Nkala et al., 2024). Collaborations with industries and organisations actively engaged in poverty alleviation projects provide a holistic view of the practical applications of STEM skills (Nkala et al., 2024). Collaborating with organisations involved in community development projects allows students to contribute to initiatives such as building schools, improving water and sanitation facilities, or creating sustainable agricultural practices. According to Nani (2016), in Zimbabwe, instead of School Development Councils (SDCs) solely managing business projects in schools, they should involve students as well. They can begin by conducting entrepreneurial projects with students and eventually hand over these projects for the students to run. Students can use afternoons for entrepreneurial activities (Nani, 2016). This can start as a pilot project and then be expanded on a larger scale once the desired results are evident. To achieve this, it is essential to involve various stakeholders, including educationists and policymakers (Saiden, 2017).

Engaging students in hands-on experiences outside the classroom is crucial for connecting theoretical knowledge to real-world challenges in Zimbabwe (Nkala et al., 2024). Students should also be exposed to the real world of business by participating in field trips to industrial areas, attending agricultural shows, and visiting business expos (Madusise, 2019). Taking students on field trips to relevant sites, such as renewable energy facilities, healthcare clinics, or community development projects, provides firsthand exposure to the impact of STEM in addressing poverty (Nani, 2016; Madusise, 2019). Structuring the curriculum around problem-based learning enables students to collaborate on real-world projects, fostering teamwork and problem-solving.

As part of its STEM strengthening strategies, the Ministry of Higher and Tertiary Education, Science and Technology Development has established Industrial and Technology Parks and Innovation Hubs in some universities in Zimbabwe. These initiatives are expected to promote developing and commercialising innovative projects and emerging technologies relevant to Zimbabwe's socio-economic development while remaining rooted in the country's heritage philosophy.

Experiential learning allows students to understand the context in which their STEM skills can be applied to address poverty-related challenges. Experiential learning enhances students' readiness for STEM careers by providing practical experiences beyond theoretical knowledge. Students witness the direct impact of their STEM contributions on individuals and communities, fostering a sense of social responsibility. Integrating experiential learning opportunities into the STEM curriculum is a transformative step in using STEM education as a praxis to end poverty. By providing hands-on experiences, internships, apprenticeships, and collaborations with industries and organisations, students enhance their technical skills and develop a profound understanding of how STEM can contribute to meaningful social change. This approach positions STEM education as a dynamic force for addressing real-world challenges and preparing students to actively participate in poverty alleviation activities.

Curriculum Transformation in STEM Education: Global Perspectives

In a curriculum turn aimed at utilising STEM education as a praxis to end poverty, the integration of global perspectives emerges as a crucial element. Incorporating global perspectives into the curriculum is essential to broaden students' understanding of poverty as a global issue. This global outlook helps students appreciate the interconnectedness of the world and the potential for STEM solutions to have a global impact on poverty. Poverty is a complex, multifaceted issue with global implications. A curriculum that incorporates global perspectives equips students with a broader understanding of poverty, emphasising its interconnectedness with various social, economic, and environmental factors (Gadzirayi et al., 2016). It fosters empathy, cultural awareness, and a sense of shared responsibility in addressing global challenges. A curriculum turn recognizes the need to integrate global perspectives seamlessly into STEM education, allowing students to engage with global challenges related to poverty, such as access to education, healthcare, clean water, and sustainable development (Nezandonyi, 2023). This exploration encourages them to view these challenges as interconnected issues that require collaborative, cross-border solutions.

Facilitating opportunities for international collaborations exposes students to a variety of perspectives and approaches to STEM challenges (Kumba et al., 2024). Collaborative projects with students from different parts of the world encourage teamwork, cross-cultural communication, and the sharing of diverse problem-solving strategies (Kumba et al., 2024). Incorporating global case studies into the curriculum provides real-world examples of successful STEM initiatives addressing poverty (Ngwenya et al., 2017). Students analyse these cases to understand the impact of science, technology, engineering, and mathematics on global development and poverty alleviation. Through its Department of Scholarships, the Zimbabwean government continues to provide numerous STEM scholarships to underprivileged students for studies at international universities under the Presidential Scholarships scheme (Zvavahera, 2014). Additionally, the government of Zimbabwe has implemented various programs and projects to enhance the quality of STEM teaching and learning. One significant initiative was the Zimbabwe-Cuba program, which began in 1986 (Gadzirayi et al., 2016). This program successfully trained over 2000 science and mathematics teachers throughout its duration. Furthermore, in 2016, the Ministry provided an opportunity for ten Lower Sixth STEM students to win a trip to the United States to tour Microsoft and other companies in Silicon Valley (Gambanga, 2016).

Highlighting successful STEM education programs with a global focus serves as inspiration for educators and students. Participation in international STEM competitions exposes students to global challenges and allows them to showcase their problem-solving skills on an international stage (Gadzirayi et al., 2016). Competitions like the International Science Olympiad foster collaboration and friendly competition among students from diverse backgrounds (Mandiudza, 2021). Programs facilitating exchanges between students from different countries provide firsthand experiences in understanding global perspectives. Exchanges may include collaborative research projects, joint seminars, and cultural exchange activities.

Virtual platforms that connect students globally for collaborative STEM projects create opportunities for ongoing engagement (Wang & Shen, 2023). These platforms foster a sense of global community enabling students to work together on shared challenges, transcending geographical boundaries. Collaborations with international organisations focused on STEM education and poverty alleviation bring valuable resources and expertise into the curriculum. These partnerships may involve joint initiatives, mentorship programs, and access to global networks (Kumba et al., 2024; Makonye & Moodley, 2023).

Incorporating global perspectives into the STEM curriculum is pivotal in a curriculum turn. By broadening students' horizons, fostering global awareness, and showcasing successful global STEM initiatives, STEM education becomes a transformative force for preparing students to address poverty on a global scale. This approach enhances the educational experience and positions STEM as a praxis that transcends borders, contributing to a more equitable and interconnected world.

Addressing these issues requires a paradigm shift in teacher training programs towards promoting interdisciplinary approaches, fostering collaboration among educators from different STEM disciplines, and emphasising the application of knowledge in real-world contexts (Mutseekwa, 2021). By incorporating innovative pedagogical strategies that promote integration and situated learning, teacher training institutions can better prepare educators to inspire students' interest and engagement in STEM fields (Kadziya & Ndebele, 2020). It is crucial to provide ongoing professional development opportunities for teachers to enhance their digital literacy skills (Dzinoreva & Mavunga, 2022). Empowered teachers can better guide students in navigating the digital landscape and integrating technology effectively into their lessons (Dzinoreva & Mavunga, 2022).

In conclusion, a curriculum turn in STEM education that places poverty alleviation at its core can serve as a transformative praxis. By preparing students to apply STEM skills in addressing real-world challenges, fostering ethical awareness, and promoting inclusivity, this approach empowers the next generation of STEM practitioners to make meaningful contributions to the global effort to eradicate poverty.

References

Alkair, S., Ali, R., Abouhashem, A., Aledamat, R., Bhadra, J., Ahmad, Z., Sellami, A., & Al-Thani, N. J. (2023). A STEM model for engaging students in environmental sustainability programs through a problem-solving approach. *Applied Environmental Education and Communication, 22*(1), 13–26. https://doi.org/10.1080/1533015X.2023.2179556

Attard, C., Berger, N., & Mackenzie, E. (2021). The positive influence of inquiry-based learning teacher professional learning and industry partnerships on student engagement with STEM. *Frontiers in Education, 6*(693221). https://doi.org/10.3389/feduc.2021.693221

Baterna, H. B., Mina, T. D. G., & Rogayan, D. V. (2020). Digital literacy of STEM senior high school students: Basis for enhancement program. *International Journal of Technology in Education, 3*(2), 105. https://doi.org/10.46328/ijte.v3i2.28

Bybee, R. W. (2010). Advancing STEM education: A 2020 vision. *Technology and Engineering Teacher, 70*, 30. Retrieved from https://search.proquest.com/openview/75bbe8b13bf3f54ebd755333ffd8621e/1?pq-origsite=gscholar&cbl=34845

Chingarande, S. D., Muchabaiwa, W., & Muyambo, T. (2023). *Bridging the gender equality gap in science at the women's university in Africa, Zimbabwe.* Final Technical Report submitted to IDRC.

Chiremba, E. (2023). The adverse influencing factors to programs and policies on education access and opportunities in Zimbabwe: A literature review. *North American Academic Research, 6*(12), 93–104. https://doi.org/10.5281/zenodo.10452297

Chirinda, B., Vere, J., & Sunzuma, G. (2023). An overview of the Zimbabwean education system. In B. Chirinda, L. Sibanda, J. Vere, & G. Sunzuma (Eds.), *Science, Mathematics and Technology education in Zimbabwe: Research, policy and practice* (pp. 159–189). Peter Lang.

Chirume, E., & Kwabena, D. (2016). Toward a better understanding of the access to education through online learning in the case of Zimbabwe. *International Journal of Advanced Research, 4*(7), 1304–1313. https://doi.org/10.21474/ijar01/1031

Chitate, H. (2016). Science, Technology, Engineering and Mathematics (STEM): A case study of Zimbabwe's educational approach to industrialisation. *World Journal of Education, 6*, 27–35.

Dekeza, C., & Kufakunesu, M. (2017). Implementation of STEM curriculum in rural secondary schools in Zimbabwe: Limits and possibilities. *Journal of Emerging Trends in Educational Research and Policy Studies, 8*(1), 11–15.

Dube, P. S. (2018). An investigation of the Science Technology Engineering and Mathematics (STEM) initiatives in the Zimbabwean education system. *International Journal for Cross-Disciplinary Subjects in Education (IJCDSE), 9*(1), 3329–3334.

Dube, T., Ncube, S. B., Mapuvire, C. C., Ndlovu, S., Ncube, C., & Mlotshwa, S. (2021). Interventions to reduce the exclusion of children with disabilities from education: A Zimbabwean perspective from the field. *Cogent Social Societies, 7*(1), 1–18.

Dzinoreva, T., & Mavunga, G. (2022). Integrating ICTs into the Zimbabwean secondary school pre-service teachers' curriculum. *Journal of Education, 88*, 53–68.

Foi, L. Y., & Kean, T. H. (2022). STEM education in Malaysia: An organisational development approach? *International Journal of Advanced Research in Future Ready Learning and Education, 29*(1), 1–19.

Gadzirayi, C. T., Bongo, P. P., Ruyimbe, B., Bhukuvhani, C., & Mucheri, T. (2016). *Diagnostic study on status of STEM in Zimbabwe*. Bindura University of Science Education and Higher life Foundation.

Gambanga, N. (2016). STEM students at public schools in Zimbabwe offered free education. *Techzim*. Retrieved from http://www.techzim.co.zw/2016/01/stem-students-publicschools-zimbabwe-offered-free-education/

Garwe, E. C., & Chikwiri, E. (2021). De-gendering STEM: Best practices from a case study of a Women's University in Africa. *Global Scientific Journal, 9*(1), 576–593.

Gondo, T., & Mudekunye, J. J. (2020). Nurturing curiosity learning through STEM in physical education in Zimbabwe. *International Journal of Technology-Enabled Student Support Services, 10*(2), 20–30.

Gory, D., Bhatia, J., & Reddy, V. R. M. (2021). From content knowledge to competencies and exams to exit profiles: Education reform in Zimbabwe. In F. M. Reimers (Ed.), *Implementing deeper learning and 21st century education reform* (pp. 145–169). Springer. https://doi.org/10.1007/978-3-030-57039-2_7

Gudyanga, A., Gudyanga, E., & Mutemeri, J. (2013). Innovative Science trends that have occurred in Zimbabwe. *Asian Social Science, 9*(12), 50–56. https://doi.org/10.5539/ass.v9n11p50

Gweshe, G. T., & Chiware, M. (2023). Gender-inclusive education in Science, Technology, Engineering, and Mathematics (STEM) fields in postindependence Zimbabwe. In D. Mhlanga & E. Ndhlovu (Eds.), *Economic inclusion in post-independence Africa. Advances in African economic, social and political development*. Springer. https://doi.org/10.1007/978-3-031-31431-5_16

Hrynevych, L., Khoruzha, L., Rudenko, N. & Proshkin, V. (2022). STEM education in the context of improving the science and mathematics literacy of pupils. *Journal of Physics: Conference Series, 2288*, 12031. https://doi.org/10.1088/1742-6596/2288/1/012031

Isaacs, S. (2007). *ICT in education in Zimbabwe*. InfoDev.

Kadziya, L., & Ndebele, C. (2020). An analysis of the challenges faced by teachers in the delivery of Science, Technology, Engineering and Mathematics education in rural day secondary schools in Zimbabwe. *International Academic Research Journal of Engineering and Technology, 1*(3), 85–95.

Kalolo, J. F. (2016). Re-aligning approaches for successful implementation of STEM education in today's elementary schools in developing countries: Policy commitments and practices. *Journal of Education and Literature, Research Academy of Social Sciences, 4*(2), 61–76.

Karimi, H., & Pina, A. (2021). Strategically addressing the soft skills gap among STEM undergraduates. *Journal of Research in STEM Education, 7*(1), 21–46. https://doi.org/10.51355/jstem.2021.99

Kumba, H., Olanrewaju, O., Akpan, J., & Mafuratidze, F. (2024). *Engineering education and industry synergy in Zimbabwe: A divide and redirection*. IntechOpen.

Mabhanda, W. (2016). Opportunities and factors affecting adoption of STEM education: The case of Gweru Polytechnic first year commerce students. *International Journal of Business Marketing and Management, 1*(5), 1–8.

Madusise, S. (2019). Possibilities for implementation of STEM education in Zimbabwean underresourced mathematics classrooms. *Journal of African Interdisciplinary Studies, 3*(7), 110–124.

Makonye, J. P., & Moodley, N. P. (2023). Connecting mathematics to STEM education: Interdisciplinary teaching and learning facilitation. *ZDM Mathematics Education, 55*, 1365–1373. https://doi.org/10.1007/s11858-023-01522-2

Mandiudza, M. (2021). Factors motivating female students from Masvingo, Zimbabwe—The choice of taking natural sciences at university. *European Journal of Social Sciences Studies, 6*(2). https://doi.org/10.46827/ejsss.v6i2.1001

Manokore, K., & Sibanda, L. (2024). National STEM education framework: Teachers' perspectives on the 2015–2022 curriculum cycle. *Journal of Research in Education and Pedagogy, 1*(2), 98–106. https://doi.org/10.70232/jrep.v1i2.12

Mapfumo, S. (2015). *The new curriculum framework: 2015–2022*. Document (unpublished). CDTS.

Maphosa, V., Dube, B., & Jita, T. (2021). Sustainable information and learning access at a rural university in Zimbabwe through a mobile application. *International Journal of Information and Education Technology, 11*(2), 82–87. https://doi.org/10.18178/ijiet.2021.11.2.1493

Marazanye, B. (2024). *The importance of play-based learning in Early Childhood Education: Selected case studies between Swedish and Zimbabwean pre-schools* (Dissertation). Retrieved from https://urn.kb.se/resolve?urn=urn:nbn:se:mau:diva-68557

Mashoko, D. (2022). Indigenous artefacts and physics curriculum: Teaching science as a cultural way of knowing. *Culture Studies of Science Education, 17*, 863–874. https://doi.org/10.1007/s11422-021-10085-1

Masunda, M. M., Chitumba, C., Mushayavanhu, T. P., & Simuka, J. (2018). An investigation of the profiles of Zimbabwean stem undergraduate freshmen as input to entrepreneurship education for stem students. *Journal of Language, Technology and Entrepreneurship in Africa, 9*, 69–89.

Matindike, F., & Ramdhany, V. (2024). Incorporating indigenous knowledge perspectives in integrated STEM education: A systematic review. *Research in Science and Technological Education*, 1–21. https://doi.org/10.1080/02635143.2024.2413675

Ministry of Primary and Secondary Education. *Curriculum review process narrative report 2014–2015*. Harare.

Ministry of Primary and Secondary Education. *Zimbabwe education blueprint: 2015–2022*.

Ministry of Primary and Secondary Education. (2015a). *Curriculum framework for primary and secondary education 2015–2022*. Zimbabwe Ministry of Primary and Secondary Education.

Ministry of Primary and Secondary Education. (2016). *Annual report*. MoPSE.

Moyo, M. (2019). Towards implementing a STEMitised curriculum in inclusive secondary school contexts: The case of five selected schools in Mzilikazi District of Bulawayo. *Zimbabwe Journal of Educational Research, 31*(3), 470-484.

Moyo, Z. (2022). The fourth Industrial revolution: A literature study of challenges associated with access to education in rural schools in Zimbabwe. *Journal of Educational and Social Research, 12*, 125–136.

Moyo, P. V., & Berdut, I. R. (2017). Interrogating the implementation of STEM education in Zimbabwe. *The International Journal of Humanities and Social Studies, 5*(11), 332–337.

Mpofu, V. (2019). A theoretical framework for implementing STEM education. In K. G. Fomunyam (Ed.), *Theorizing STEM education in the 21st century*. IntechOpen. isbn:978-1-78985-702-3.

Mpofu, V., & Mutseekwa, C. (2023). School science in Zimbabwe: Debate and critique for new directions. In B. Chirinda, L. Sibanda, J. Vere, & G. Sunzuma (Eds.), *Science, Mathematics and Technology education in Zimbabwe: Research, policy and practice* (pp. 45–66). Peter Lang.

Mugweni, R. M. (2016). Early childhood (ECD) teachers' conceptions and implementation of the child-centred approach to teaching science. *International Journal of Academic Research in Progressive Education and Development, 5*(4), 2226–6348.

Mutseekwa, C. (2021). STEM practices in Science teacher education curriculum: Perspectives from two secondary school teachers' colleges in Zimbabwe. *Journal of Research in Science, Mathematics and Technology Education, 4*, 75–92. https://doi.org/10.31756/jrsmte.422

Nani, G. V. (2016). Entrepreneurial education in the school curriculum: In search of positioning in Zimbabwe. *Problems and Perspectives in Management, 14*(3), 85–90. https://doi.org/10.21511/ppm.14(3).2016.08

Nezandonyi, J. (2023). Embedding education for sustainable development in the teaching of technology courses in Zimbabwe: Policy and practice. In B. Chirinda, L. Sibanda, J. Vere, & G. Sunzuma (Eds.), *Science, Mathematics and Technology education in Zimbabwe: Research, policy and practice* (pp. 159–189). Peter Lang.

Nguyen, T. P. L., Nguyen, T. H., & Tran, T. K. (2020). STEM education in secondary schools: Teachers' perspective towards sustainable development. *Sustainability, 12*(21), 8865. https://doi.org/10.3390/su12218865

Ngwenya, B., Pelser, T., & Sibanda, R. (2017). Success as the knowledge economy the success or failure of the Zimbabwe's STEM initiative. In *International conference on intellectual capital and knowledge management and organisational learning* (pp. 169–178). Academic Conferences International.

Nherera, C. M. (2015). The rising demand for higher education: The case of the Women's University in Africa. *Zimbabwe Journal of Educational Research, 27*(2), 221–239.

Nkala, B., Sibanda, V. M., Ndhlovu, J., Hobane, L., & Singh, S. (2024). Challenges and drivers of industrialist propensity among chemical engineering students in STEM institution in Zimbabwe: Towards a conceptual framework. *International Journal of Environment, Engineering and Education, 6*(1), 26–37. https://doi.org/10.55151/ijeedu.v6i1.126

Pakombwele, A. (2024). Digital education for sustainable development in early childhood classrooms of Zimbabwe. *Gen-Multidisciplinary Journal of Sustainable Development, 2*(2), 41–56. Retrieved from https://gmjsd.org/journal/index.php/gmjsd/article/view/53

Pakombwele, A., & Tsakeni, M. (2022). The teaching of science process skills in early childhood development classrooms. *Universal Journal of Educational Research, 10*(4), 273–280. https://doi.org/10.13189/ujer.2022.100402

Palid, O., Cashdollar, S., Deangelo, S., Chu, C., & Bates, M. (2023). Inclusion in practice: A systematic review of diversity-focused STEM programming in the United States. *International Journal of STEM Education, 10*(1), 2. https://doi.org/10.1186/s40594-022-00387-3

Parawira, W. (2016). Science in our natural world. *The Sunday Mail*. Retrieved from www.Sundaymail.co.zw

Risiro, J. (2014). An evaluation on the implementation of environmental education program at Mutare Teachers' College in Zimbabwe. *International Journal of Innovation, Research and Development, 3*(1), 264–271.

Saiden, T. (2017). Towards an entrepreneurship and STEM Education primary school curriculum in Zimbabwe: A case study of Bumburwi of Gweru District. *Advances in Social Sciences Research Journal, 4*(18), 148–159.

UNICEF. (2020). *Zimbabwe's proposal to GPE for COVID-19 accelerated funding to support the education sector*. Global Partnership for Education.

Wang, C., & Shen, J. (2023). Technology-enhanced collaborative learning in STEM. In *International encyclopedia of education* (4th ed., pp. 207–214). Elsevier. https://doi.org/10.1016/B978-0-12-818630-5.13005-2

Zambaza, G. (2022). *E-Learning platforms in Zimbabwe*. Retrieved from How To's & Guides: www.paynow.co.zwblog

Zhou, M. (2017). *STEM implementation in rural secondary schools: A case of Kwekwe Zhombe district.* Undergraduate research project, Midlands State University.

Zvavahera, P. (2014). The presidential scholarship programme in Zimbabwe: A living case of the political will in promoting regionalisation in higher education. *Journal of Case Studies in Education, 6*, 1–9.

The Future of STEM Education in a Developing Country: Re-Aligning Approaches for Successful Implementation

Abstract The Fourth Industrial Revolution (4IR), characterised by disruptive technologies, is a chief international driver of change that has implications for STEM education. Expectations of stakeholders regarding STEM output, emerging needs and constant changes are also drivers for effective STEM education. Globally, governments and industries have prioritised improving the quality of STEM learning to meet industrial and workforce needs in the near future. Zimbabwe is one of the developing countries that has made an effort to improve STEM learning because it is vital to develop the skills required to meet future demands. In light of this, in Zimbabwe, efforts have been made to enrol more students in STEM subjects so they can have various STEM professions in the future. Such futuristic predictions have implications for the STEM approaches, which must be aligned with STEM and twenty-first-century skills in the 4IR era. In developing countries, the trajectory of STEM education is critical for sustainable development and global competitiveness. This chapter explores the opportunities for reshaping STEM education strategies to align with developing nations' unique needs and contexts. It examines the impact of globalisation, technological advancements, and socio-economic factors on educational policies and practices. By synthesising existing literature, the chapter highlights innovative approaches and best practices for enhancing STEM education in developing contexts. Key considerations include policy reforms, fostering inclusive and equitable access to STEM education, investment in infrastructure and resources, and integrating emerging technologies and professional development. Through strategic alignment and collaboration among stakeholders, this chapter proposes actionable recommendations to advance the future of STEM education in developing countries, ultimately contributing to socio-economic growth, scientific advancement, and global sustainability.

B. Chirinda et al., *STEM Education in Poverty*,
https://doi.org/10.1007/978-3-031-92490-3_9

Introduction

In the rapidly evolving society, the significance of STEM knowledge and skills is consistently expanding in Zimbabwe. There is a broad consensus that addressing contemporary global challenges necessitates a multidisciplinary scientific workforce equipped with both new technologies and interdisciplinary thinking (Nkala et al., 2024). This workforce may need to integrate multiple STEM concepts to effectively tackle these challenges. Moreover, there is a widespread acknowledgment of the urgent need to cultivate and prepare a diverse STEM-literate workforce capable of comprehending and navigating the complexities of the technological landscape (Nkala et al., 2024). In the global educational landscape, STEM fields play a pivotal role in driving innovation, economic growth, and societal progress (Mabhanda, 2016).

In Zimbabwe, a robust foundation in STEM principles is now indispensable in the job market, not only within STEM-specific domains but also across diverse career sectors (Madusise, 2019). Regrettably, in Zimbabwe, disparities persist in access, participation, and achievement in STEM subjects, influenced by cultural, social, economic, gender, and geographical factors (Mabhanda, 2016). These inequities in STEM education pose a threat to addressing literacy and poverty gaps, meeting the needs of a technology-oriented market, ensuring national security, and sustaining leadership in scientific research and development (Chingarande et al., 2023). Consequently, effective STEM education in Zimbabwe is imperative to cultivate essential skills and mindsets in students from early educational stages, promote cultural inclusivity, and employ problem-solving and inquiry-based methodologies (Mabhanda, 2016).

Many developed countries and economies have recognised the strategic importance of STEM education and have implemented systematic policies to meet the demand for a highly skilled workforce elsewhere and in Zimbabwe specifically (Mutseekwa, 2021; Moyo & Berdut, 2017; Mabhanda, 2016). While developed nations have long recognised the importance of fostering robust STEM ecosystems (Aslam et al., 2023), many developing countries, including Zimbabwe, face significant challenges in this endeavour (Mabhanda, 2016; Moyo & Berdut, 2017). A STEM-learning ecosystem encompasses various components such as structures, networks, partnerships, resources, and public policies that facilitate and reinforce STEM learning (Nkala et al., 2024; Svodziwa, 2021; Kumba et al., 2024). This chapter goes beyond enhancing curricula; it aims to orchestrate comprehensive reforms aligning educational systems with twenty-first-century demands. Cultivating a STEM-savvy populace in developing countries is complex, requiring an understanding of local contexts, socioeconomic realities, and infrastructural constraints. It involves transitioning from traditional pedagogical models to dynamic, experiential learning frameworks, empowering students as active participants in the scientific and technological landscape.

The chapter elucidates pivotal components of effective STEM education, including the role of policy frameworks in catalysing systemic change, the imperative of cultivating a diverse and inclusive STEM workforce, and the integration of emerging technologies to bridge educational divides. It emphasises the significance of forging strategic partnerships between governmental bodies, educational institutions, industry stakeholders, and civil society organisations to leverage collective expertise, resources, and experiences (Kumba et al., 2024). In this era dominated by technology, recent advancements in information and communication technology are impacting all facets of society, including education, and equipping future generations with STEM proficiency is essential (Yingi et al., 2022). The chapter seeks to chart a course towards equitable, sustainable, and inclusive educational futures. It envisions a world where every individual has the opportunity to contribute meaningfully to societal advancement through STEM education.

Policy Reforms

STEM education is a multifaceted approach to education that integrates the disciplines of science, technology, engineering, and mathematics. STEM education has garnered significant attention worldwide in recent decades, emerging as a crucial component of educational reforms in many countries including Zimbabwe (Mabhanda, 2016). Policy reforms play a crucial role in shaping the future of STEM education in developing countries such as Zimbabwe (Moyo & Berdut, 2017). The government of Zimbabwe has recognized the significance of STEM fields in driving economic growth, innovation, and societal progress. It has also made some effort to allocate sufficient resources to support these STEM fields' development. STEM education is recognised as an innovative approach within global educational policy and reforms, addressing the need for competency in mathematics and science while emphasising the integration of technology and engineering in Zimbabwe (Chitate, 2016). STEM education acknowledges that modern citizens must understand science and mathematics and how these disciplines intersect with technology and engineering in today's rapidly evolving world (Gadzirayi et al., 2016). Comprehensive policy reforms should encompass various aspects, including curriculum modernisation, teacher training, and the integration of technology into educational practices (Mutseekwa, 2021; Chisom et al., 2024).

Outdated curricula often fail to keep pace with the rapidly evolving landscape of STEM disciplines, as noted by Mutsvangwa and Zezekwa (2021). Policy reforms should prioritise the revision and modernisation of curricula to ensure they are aligned with current industry trends and global standards (Kumba et al., 2024). This involves updating course content, incorporating interdisciplinary approaches, and emphasizing practical, hands-on learning experiences (Chisom et al., 2024). In developing nations, for example, in Zimbabwe, educational policies that prioritise and promote STEM in schools and teacher education curricula have been put in place (Mutseekwa, 2021). Curricula reform must take contextual factors into account and promote meaningful learning. Specifically, STEM education reviews should incorporate the rich knowledge offered by the indigenous knowledge systems of various Zimbabwean cultures (Mutsvangwa & Zezekwa, 2021). Promoting these Indigenous Knowledge Systems in STEM education should be a central focus of future curriculum reviews if Zimbabwe aims to advance its citizens.

In Zimbabwe, interdisciplinary stands as a foundational characteristic within STEM education (Parawira, 2016). Researchers underscore the importance of crafting STEM-integrated courses within curriculum design, emphasising the utilisation of interdisciplinary knowledge to address authentic, real-world problems (Madusise, 2019). Moyo and Berdut (2017) stress that students are anticipated to adeptly integrate knowledge and effectively engage in problem-solving endeavours. With interdisciplinary principles serving as the bedrock, scholars actively endeavor to eradicate disciplinary silos and forge connections within teaching methodologies (Zhou, 2017). For instance, Mpofu (2019) advocates for interdisciplinary research to underpin knowledge acquisition and ensure consistency across teaching practices, curriculum design, and assessment methods. This trajectory aligns with the global movement towards integrative STEM education, mirroring the trends of interdisciplinary approaches and problem-based learning contexts delineated within the decentralized-unified continuum of STEM education, as outlined by Nadelson and Seifert (2017). Fostering interdisciplinary learning approaches enables students to engage deeply with complex issues, empowering them to actively participate in society and enhance their overall quality of life through their contributions (Parawira, 2016; Madusise, 2019).

Researchers explore various approaches to delivering STEM education in Zimbabwe (Gadzirayi et al., 2016; Chauraya, 2023). Characteristics of teaching and learning methods in the competence-based curriculum include being participatory, research-driven, collaborative, integrated, and interdisciplinary, and inquiry-based (Ministry of Primary and Secondary Education, 2014; Chauraya, 2023). These methods include discovery, project-based learning, problem-solving, guided discovery, discussion, modelling, educational tours and design-based learning. In Zimbabwe, the new curriculum, particularly in science learning areas, introduced some practical tasks and coursework projects, which are very welcome to promote and enhance authentic learning (Bhukuvhani et al., 2023). Project-based learning offers the advantage of simulating expert research processes and emphasising the practical application of acquired knowledge in Zimbabwe (Chitate, 2016; Mpofu, 2019) in Zimbabwe. By fostering a relevant and engaging curriculum, students are better equipped to develop critical thinking, problem-solving, and innovation skills essential for success in STEM fields in the twenty-first century (Yingi et al., 2022). STEM education should not only bolster students' proficiency in innovation, creativity, scientific thinking, and problem-solving skills but also cultivate their sustainability mindsets, equipping them to tackle the multifaceted economic, social, and environmental challenges inherent in both formal and informal curricula (Yingi et al., 2022).

In light of the escalating emphasis on environmental sustainability, the future trajectory of STEM education must intertwine ecological principles, sustainable practices, and environmental stewardship in Zimbabwe (Moyo & Ndlovu, 2023). This imperative entails integrating real-world environmental challenges into curricula, prompting students to devise solutions to pressing issues such as climate change, pollution, and resource depletion (Shava, 2021; Yingi et al., 2022). STEM education holds the potential to cultivate environmentally conscious citizens and professionals who actively contribute to sustainable innovations (Chitate, 2016). Moreover, as technological advancements continue, ethical considerations assume heightened importance across STEM fields. Thus, future directions in STEM education should underscore the significance of moral reasoning, responsible innovation, and an understanding of the societal implications of technological progress. Educators play a pivotal role in instilling a sense of ethical responsibility in students, guiding them to contemplate their work's social, cultural, and environmental ramifications (Nezandonyi, 2023). By embedding ethical considerations into STEM curricula, we can ensure that the next generation of STEM professionals approaches their endeavors with a holistic comprehension of their impact.

The effectiveness of STEM education hinges on the quality of teaching. Therefore, policy reforms should focus on enhancing the skills and competencies of STEM educators through targeted teacher training and capacity-building programs (Svodziwa, 2021) in Zimbabwe. These initiatives should provide educators with opportunities for professional development, exposure to innovative pedagogical practices, and specialized training in STEM subjects, as highlighted by Mwembe et al. (2022) in Zimbabwe. By equipping teachers with the necessary tools and knowledge, they can effectively engage students and nurture their interest and proficiency in STEM disciplines. Additionally, there is a need to identify policies and strategies that will promote gender equality in STEM fields (Gweshe & Chiware, 2023).

Before the implementation of the STEM policy, scientific and research education in Zimbabwe was primarily guided by the Science and Technology Policy (Shizha & Kariwo, 2011). The main goal of this policy was to promote national scientific and technological self-reliance by ensuring rapid and sustainable industrialization through research and development. However, there is currently a noticeable policy discord between the two ministries of education. The Ministry of Higher and Tertiary Education emphasizes Science, Technology, Engineering, and Mathematics (STEM), while the Ministry of Primary and Secondary Education emphasizes STEM as well as Arts, Heritage Studies, and commercial subjects. This discord is detrimental to achieving the goals of STEM education and the Fourth Industrial Revolution (4IR), as the key policy drivers are not aligned (Yingi et al., 2022; Chitate, 2016).

Integrating Emerging Technologies

Promoting inclusive, equitable, and quality education through 4IR technologies in Zimbabwe aligns with SDG Goal number four (Yingi et al., 2022). Implementing 4IR technologies enables self-organising education, where technology plays a major role in instructional delivery and learner support (Mhlanga, 2024). This shift will drive education towards individualization and learner-centered approaches, utilizing technologies such as artificial intelligence (AI), learning analytics, and the Internet of Things (IoT) (Yingi et al., 2022). The updated curriculum for 2015 to 2022 emphasizes ICT and learner-centric learning, moving away from traditional teacher-centered methods (Ministry of Primary and Secondary Education, 2015a, 2015b; Dzinotyiweyi & Taddese, 2020). This transformation aims to produce competent, problem-solving, and critical thinkers eager to modernize society (Ministry of Primary and Secondary Education, 2015a, 2015b).

The STEM education of the future will leverage technology in innovative ways to provide equitable access to all learners and foster environments where every student can thrive. Skilful instruction supported by technological advancements has the potential to overcome structural barriers such as cost, distance, opportunity, socioeconomic background, or prior STEM preparation, ultimately enabling all STEM learners to overcome stereotypes and biases with the support of their learning communities (Dzinotyiweyi & Taddese, 2020).

Technology plays a transformative role in enhancing the quality and accessibility of education. Policy reforms should prioritize the integration of technology into educational practices to enrich STEM learning experiences (Dzinoreva & Mavunga, 2022). This involves equipping schools with digital resources, such as computers, tablets, and interactive learning tools, and leveraging online platforms for content delivery and collaboration. Additionally, initiatives to promote digital literacy among students and educators are essential for harnessing the full potential of technology in STEM education (Chimhenga, 2017). By embracing technology-enabled teaching and learning approaches, schools can create dynamic and interactive environments that inspire curiosity and facilitate deeper understanding of STEM concepts.

In Zimbabwe, the future of STEM education is poised for transformation through the integration of emerging technologies, which hold the potential to revolutionize teaching and learning of STEM subjects (Yingi et al., 2022). As stated in chapter "STEM Education Pedagogy in Developing Country Contexts: Relevance and Challenges", VR and AR provide immersive experiences that can transport students to otherwise inaccessible places, offering them unprecedented opportunities to explore complex concepts. AI stands to personalize learning experiences by adapting to individual student needs and offering targeted (Yingi et al., 2022). When technology is effectively integrated into STEM education, it has the capacity to boost student engagement in STEM learning and elevate the overall quality of STEM education (Chimhenga, 2017). Additionally, gamification, employing game elements in non-game contexts, can enhance engagement and motivation among students. The continued advancement of these technologies, as highlighted by the Zimbabwe Flying Labs (2020), suggests a promising future for STEM education characterised by dynamic, interactive, and personalised learning environments. By leveraging these tools, educators can create innovative approaches to teaching and learning that cater to the diverse needs of students, ultimately fostering a deeper understanding and appreciation for STEM subjects.

Technology enables the creation and dissemination of digital resources and online learning platforms that offer flexible, accessible, and personalized learning experiences for students (Maune, 2023). These resources can include interactive simulations, virtual laboratories, multimedia tutorials, and adaptive learning software tailored to individual learning needs. By providing access to a diverse range of digital resources, technology empowers learners to explore STEM concepts at their own pace, regardless of geographical location or socio-economic status (Dzinotyiweyi & Taddese, 2020).

Technology facilitates remote learning and distance education, allowing students to participate in STEM courses and activities from anywhere with an internet connection. Virtual classrooms, video conferencing tools, and online collaboration platforms enable educators to deliver engaging STEM instruction and facilitate interactive discussions with students across different locations (Mhlanga, 2024). Additionally, recorded lectures and asynchronous learning materials provide flexibility for students with diverse schedules and commitments, ensuring equitable access to educational opportunities (Dzinotyiweyi & Taddese, 2020).

In Zimbabwe, the widespread availability of mobile devices such as smartphones and tablets expands access to STEM education for students who may not have access to traditional desktop computers or laptops (Gomba, 2016). Mobile learning applications, educational games, and interactive e-books offer opportunities for on-the-go learning and engagement, catering to diverse learning styles and preferences (Maphosa et al., 2021). Moreover, ensuring the accessibility of digital content and devices for students with disabilities promotes inclusivity and equity in STEM education (Dube et al., 2021).

Technology facilitates data-driven instruction and personalized learning experiences tailored to the individual needs and interests of students (Chimhenga, 2017). Learning analytics tools and intelligent tutoring systems analyze student performance data to identify areas of strength and weakness, provide targeted feedback, and recommend personalized learning pathways. By leveraging technology to adapt instruction to the unique learning profiles of students, educators can optimize learning outcomes and support the success of all learners, regardless of their prior STEM preparation or background.

In Africa, technology enables students to engage with STEM learning communities and participate in collaborative projects and discussions with peers, experts, and mentors from around the world (Baingana, 2024). Online forums, social media platforms, and virtual communities provide spaces for students to share ideas, seek support, and collaborate on STEM-related projects, fostering a sense of belonging and connection within the STEM community. Additionally, virtual internships, remote mentoring programs, and online competitions offer opportunities for students to gain real-world experience and build professional networks, regardless of their geographic location or socio-economic background.

The integration of technology into STEM education holds immense potential for providing equitable access and empowering all learners to thrive in STEM fields in Africa (Chisom et al., 2024). By leveraging digital resources, online learning platforms, mobile devices, data-driven instruction, and online collaboration tools, educators can overcome structural barriers and create inclusive learning environments where every student has the opportunity to succeed. With the support of technology and their learning communities, STEM learners can overcome stereotypes and biases, unlock their full potential, and contribute to a more diverse, innovative, and inclusive STEM workforce.

In addition to policy reforms, governments should implement incentives and support mechanisms to encourage the adoption of STEM education initiatives (Mabhanda, 2016). This may include offering scholarships, grants, and tax incentives for institutions and individuals involved in STEM education. By incentivizing innovation and investment in STEM, governments can stimulate growth in these fields and attract talented individuals to pursue careers in STEM-related industries (Mabhanda, 2016). For instance, in 2015, the Zimbabwean government introduced the idea of maths and teacher retention allowances with the aim to curtail science teacher turnover as well as encourage more teachers in the science disciplines (Gadzirayi et al., 2016). However, teachers from non-science discipline resisted the move, and it was abandoned by the government, even though it appeared to be a valuable avenue to improve the presence of science teachers in rural schools, which is worth revisiting. In addition, in 2016, the Ministry implemented several strategies to increase the number of students taking up STEM subjects at the high school level, including offering a cash prize of USD100,000 for schools that registered full Lower Sixth Form STEM classes in 2016 (Chara, 2016; Gambanga, 2016). The government and private companies, including NetOne, POTRAZ, Powertel, TelOne, Transmedia, ZARNet, and ZimPost, are promoting the integration of information technology into the curriculum in Zimbabwe (Chirume & Kwabena, 2016).

In Zimbabwe, the trajectory of STEM education's future is intricately tied to the formulation of educational policies governing curricula, funding, and the overarching direction of educational institutions (Gadzirayi et al., 2016). Policymakers are responsible for adapting to the evolving landscape of STEM education by cultivating an environment conducive to innovation (Chitate, 2016). This entails earmarking resources for teacher professional development, modernizing curricula to reflect the interdisciplinary and technological nature of STEM fields, and ensuring equitable access to state-of-the-art resources (Yingi et al., 2022). In 2016, the Ministry gave away 100 laptops and 100 iPads to lucky Lower Sixth Form STEM students (Chara, 2016; Gambanga, 2016). Furthermore, policies should actively promote collaborations between educational institutions, industry partners, and research organizations to bridge the gap between academic learning and real-world applications (Yingi et al., 2022; Kumba et al., 2024). The insights from Moyo and Berdut (2017) underscore the critical role of policy in shaping the future trajectory of

STEM education, emphasizing the need for proactive measures to meet the demands of an ever-evolving educational landscape.

Investment in Infrastructure and Resources

Investment in educational infrastructure and resources is paramount for nurturing a conducive learning environment that fosters the growth and development of students, particularly in STEM fields in Zimbabwe (Mabhanda, 2016; Mutsvangwa & Zezekwa, 2021). According to Mabhanda (2016), as a developing country, Zimbabwe must prioritize investment in resources and infrastructure in under-resourced schools to ensure equitable access to quality education for all. This investment encompasses various components, including the provision of laboratories, mobile laboratories, Information and Communication Technology (ICT) facilities, and educational materials essential for hands-on experimentation and inquiry-based learning (Mabhanda, 2016).

Well-equipped laboratories are indispensable for facilitating practical learning experiences in Zimbabwe's STEM subjects, such as biology, chemistry, physics, and engineering (Mwembe et al., 2022; Svodziwa, 2021). These spaces allow students to conduct experiments, analyse data, and develop critical thinking and problem-solving skills. Investment in laboratory infrastructure should focus on equipping schools with modern, state-of-the-art facilities that meet safety standards and accommodate hands-on exploration (Mutseekwa, 2021). Additionally, ensuring the availability of laboratory supplies and equipment is essential for enabling students to engage in meaningful scientific inquiry and exploration (Mwembe et al., 2022).

In today's digital age, access to Information and Communication Technology (ICT) is essential for enhancing learning outcomes and preparing students for future challenges inZimbabwe (Mutseekwa, 2021). Investment in ICT infrastructure should include the provision of computers, internet connectivity, multimedia resources, and educational software (Kangara et al., 2022; Dzinotyiweyi & Taddese, 2020; Dzinoreva & Mavunga, 2022). These tools empower students to access information, collaborate with peers, and engage in interactive learning experiences. Moreover, ICT facilities enable teachers to integrate technology into their instructional practices, delivering dynamic and engaging lessons that cater to diverse learning styles.

Access to high-quality educational materials is fundamental for supporting teaching and learning in STEM subjects in Zimbabwe (Svodziwa, 2021). Investment in educational materials should encompass textbooks, reference materials, laboratory manuals, and supplementary resources tailored to the curriculum. Additionally, providing schools with access to digital libraries, open educational resources (OER), and online learning platforms expands students' access to a wealth of educational content and promotes self-directed learning. Moreover, investing in developing locally relevant and culturally responsive educational materials ensures that students have access to resources that resonate with their lived experiences and promote meaningful learning outcomes.

STEM researchers, educators, and relevant government departments could devise innovative schemes for locally producing low-cost equipment and apparatus using available resources (Mutsvangwa & Zezekwa, 2021). This approach proved fairly successful in Zimbabwe immediately after its independence from Britain in the early 1980s through the Zimbabwe Secondary School Science Project (ZIM-SCI). The project enabled students in remote and rural areas to have the same scientific learning experiences as their urban counterparts, who had easier access to conventional scientific equipment (Mutsvangwa & Zezekwa, 2021). Sustained investment in infrastructure maintenance and upkeep is essential for ensuring the longevity and effectiveness of educational facilities. Regular maintenance of laboratories, ICT equipment, and educational materials is necessary to prevent deterioration and ensure optimal functionality. Additionally, providing ongoing technical support and training for teachers and staff helps maximise the utilisation of infrastructure resources and promotes a culture of innovation and excellence in education.

In conclusion, investment in educational infrastructure is a critical enabler for fostering a conducive learning environment and promoting the development of STEM education in developing countries. By prioritising the provision of laboratories, ICT facilities, and educational materials, governments can empower students with the resources they need to thrive in STEM fields and contribute to the advancement of society. Collaboration among policymakers, educators, industry stakeholders, and the community is essential for mobilising resources and implementing sustainable infrastructure development initiatives that benefit present and future generations.

Teacher Training and Professional Development

Teacher training and professional development are critical components of enhancing the quality of STEM education in developing countries, including Zimbabwe (Mwembe et al., 2022; Svodziwa, 2021). Continuous learning opportunities empower educators with the pedagogical skills, subject knowledge, and innovative teaching strategies necessary to effectively engage students and foster their interest and proficiency in STEM disciplines (Mwembe et al., 2022). Collaborative initiatives between governments, academic institutions, and private sector stakeholders play a pivotal role in facilitating knowledge exchange and capacity building within the teaching community (Mandiudza, 2021).

In Zimbabwe, implementing comprehensive professional development programs tailored to the needs of STEM teachers is essential for equipping them with the necessary skills and competencies to excel in their roles (Mwembe et al., 2022). These programs should offer a diverse range of training opportunities, including workshops, seminars, conferences, and online courses, covering topics such as inquiry-based teaching methods, STEM curriculum development, assessment strategies, and classroom management techniques (Mutsvangwa & Zezekwa, 2021). By providing educators with ongoing support and professional growth opportunities, they can stay abreast of the latest advancements in STEM education and continually enhance their teaching practices to meet the evolving needs of their students (Mwembe et al., 2022).

According to Gadzirayi et al. (2016), the Government of Zimbabwe implemented various programs and projects to enhance the quality of STEM teaching and learning. One such initiative is the establishment of the Bindura University of Science Education, which is specifically mandated to focus on training STEM educators. Initially founded in 1996 as a university college of science education under the University of Zimbabwe, it became a fully independent university in 2000. At Bindura University of Science Education the Virtual and Open Distance Learning (V.O.D.L) program emerged as an internally developed strategy to meet the training needs of prospective science educators who are currently employed or aspire to work as school teachers (Dodo et al., 2013). The University of Zimbabwe has been offering a Bachelor of Education degree in Science and Mathematics Education using the Open Distance electronic Learning (ODeL) methodology, which prioritizes flexibility and accessibility over traditional face-to-face learning. As part of this initiative, three secondary school Teachers' Colleges—Mutare, Hillside, and Belvedere Technical Teachers' College—have been designated as satellite centers for the ODeL degree program (Ruparanganda et al., 2012).

Specialised training in STEM subjects is crucial for ensuring educators have the requisite content knowledge to effectively teach these disciplines in Zimbabwe, as suggested by Mwembe et al. (2022). Collaborative initiatives between academic institutions and industry experts can facilitate subject-specific training programs that deepen educators' understanding of key STEM concepts and methodologies (Mutseekwa, 2021; Mabhanda, 2016). These programs may involve partnerships with universities, research institutions, and STEM-focused organizations to provide educators with access to cutting-edge research, resources, and expertise in their respective fields (Mutsvangwa & Zezekwa, 2021). For instance, in 2014, the Ministry of Primary and Secondary Education in Zimbabwe launched the Teacher Capacity Development (TCD) Programme, aiming to train approximately 5000 teachers in STEM by the end of 2015 (Gadzirayi et al., 2016). Programs like the TCD are seen as crucial for building teachers' confidence and competence in delivering STEM education (Mutseekwa, 2021). The TCD program was launched to address critical shortages of teachers in subjects like science and mathematics and to meet the needs of the competence-based curriculum. The TCD program allows teachers to improve their qualifications in various specialties. Universities specializing in training science teachers under this program include Bindura University of Science Education (BUSE), which offers BSc and MSc programs in Mathematics, Biology, Chemistry, and Physics; Midlands State University (MSU), which offers a BEd in Computer Science; and Zimbabwe Open University (ZOU), which offers a BEd in Science Education. Additionally, mentorship programs pairing experienced educators with novice teachers can foster knowledge transfer and support professional growth within the teaching community.

In today's rapidly advancing digital age, integrating technology into educational practices is paramount for enhancing teaching and learning experiences, particularly in STEM education (Maphosa et al., 2021). In Zimbabwe technology integration training programs play a crucial role in equipping educators with the necessary skills and confidence to effectively leverage digital tools, multimedia resources, and online platforms to enhance STEM instruction (Dzinoreva & Mavunga, 2022). These training initiatives are designed to empower educators to harness the power of technology to create engaging, interactive, and personalized learning experiences for their students. Training programs focused on technology integration equip educators with the skills and confidence to leverage digital tools, multimedia resources, and online platforms to enhance STEM instruction (Dzinoreva & Mavunga, 2022). These programs may cover topics such as digital literacy, educational software applications, virtual laboratories, and interactive multimedia resources. Collaborative partnerships with technology companies and educational organizations can provide educators with access to training workshops, online resources, and technical support to effectively integrate technology into their teaching practices (Dzinotyiweyi & Taddese, 2020).

In Zimbabwe, technology integration training programs often begin by providing educators with foundational knowledge and skills in digital literacy (Chitanana, 2015). Educators learn how to navigate digital devices, use basic software applications, and access online resources effectively (Chirume & Kwabena, 2016). By building educators' digital literacy skills, training programs lay the groundwork for more advanced technology integration efforts and ensure that educators feel comfortable and confident using digital tools in their teaching practice (Chitanana, 2015).

Technology integration training programs introduce educators to a wide range of educational software applications designed to support STEM instruction in Zimbabwe (Chitanana, 2015; Dzinotyiweyi & Taddese, 2020). These applications may include interactive simulations, virtual laboratories, coding platforms, data analysis tools, and multimedia resources tailored to STEM subjects. Educators learn to integrate these software applications into their lesson plans to enhance student engagement, facilitate active learning, and deepen understanding of STEM concepts through hands-on exploration and experimentation.

In Zimbabwe, virtual laboratories offer students the opportunity to conduct experiments and investigations in a simulated online environment, providing access to resources and equipment that may be otherwise unavailable due to cost, logistics, or safety considerations (Mpofu & Mutseekwa, 2022). In Zimbabwe, Mwembe et al. (2022) recommend mobile laboratories for under-resourced rural schools because one mobile laboratory can service many schools within a district. Technology integration training programs teach educators how to leverage virtual laboratories to provide authentic and immersive learning experiences that foster inquiry-based learning, critical thinking, and problem-solving skills in STEM subjects, such as biology, chemistry, physics, and environmental science (Mpofu & Mutseekwa, 2022).

Multimedia resources, such as interactive presentations, videos, animations, and digital simulations, can enhance STEM instruction by providing visualizations, demonstrations, and real-world examples that make abstract concepts more tangible and accessible to students. Technology integration training programs demonstrate how educators can incorporate interactive multimedia resources into their lessons to illustrate complex concepts, engage students' interest, and reinforce learning objectives (Mpofu & Mutseekwa, 2022).

In Zimbabwe, Mandiudza (2021) proposed that collaborative partnerships between technology companies, educational organizations, and academic institutions play a vital role in providing educators with access to technology integration training programs, resources, and technical support. These partnerships may involve hosting training workshops, webinars, and professional development sessions led by experts in educational technology (Mandiudza, 2021). Additionally, technology companies and educational organizations may offer online resources, tutorials, and technical support services to assist educators in effectively integrating technology into their teaching practices and addressing any technical challenges they encounter (Chirume & Kwabena, 2016).

Technology integration training programs are essential for equipping educators with the skills, knowledge, and resources needed to effectively integrate technology into STEM instruction (Baingana, 2024). By providing educators with training in digital literacy, educational software applications, virtual laboratories, interactive multimedia resources, and access to collaborative partnerships and technical support, these programs empower educators to create dynamic, engaging, and impactful learning experiences that prepare students for success in the digital age and the STEM fields of the future. The Research Council of Zimbabwe, in partnership with Endasolutions (Pvt.) Ltd., an ICT consultancy company, developed and introduced Virtual Learning Environment (VLE) software to all Government Colleges, Teacher Training Colleges, and A-Level schools. The primary objective was to increase potential enrollments in higher learning institutions and provide resource materials for prospective students who wish to continue with their education through OPS (Baingana, 2024).

Establishing a vibrant community of practice for STEM educators fosters collaboration, knowledge sharing, and professional networking opportunities, as reported in Zimbabwe by Mandiudza (2021). Collaborative initiatives between governments, academic institutions, and industry stakeholders can facilitate the creation of online forums, discussion groups, and professional learning communities where educators can exchange ideas, share best practices, and seek support from their peers (Mandiudza, 2021). Additionally, mentorship programs, teacher networks, and collaborative projects can promote collaboration and collective learning among educators, enriching their teaching practices and enhancing student outcomes.

Recognizing and rewarding excellence in teaching through incentives such as awards, grants, and professional certifications motivates educators to pursue continuous professional development and strive for excellence in their practice (Kalolo, 2016). Collaborative initiatives between governments, educational institutions, and private sector stakeholders can establish mechanisms for recognizing outstanding STEM educators and providing them with opportunities for career advancement and professional growth (Kumba et al., 2024; Botha, 2022). By valuing and investing in the professional development of educators, stakeholders can cultivate a culture of excellence and innovation in STEM education, ultimately benefiting students and society as a whole (Kumba et al., 2024).

Continuous professional development programs are essential for enhancing the pedagogical skills of STEM educators and promoting excellence in STEM education in developing countries. Collaborative initiatives among governments, academic institutions, and private sector stakeholders are vital in facilitating knowledge exchange, capacity building, and professional growth within the teaching community. By investing in teacher training and professional development, stakeholders can empower educators with the tools and support they need to inspire the next generation of STEM leaders and innovators.

Promoting Inclusivity and Diversity

Promoting inclusivity and diversity within STEM fields is a matter of equity and a strategic imperative for driving innovation and addressing complex global challenges (Yingi et al., 2022). Efforts to foster inclusivity and diversity within STEM education and professions are essential for addressing gender and socio-economic disparities, creating opportunities for underrepresented groups, and cultivating a more diverse talent pool in STEM disciplines (Yingi et al., 2022). Collaborative initiatives between governments, educational institutions, industry partners, and community organizations play a crucial role in promoting inclusivity and diversity within STEM fields (Botha, 2022).

In Zimbabwe, the country's Vision 2030 underscores the critical role of science and technology in the nation's industrialization agenda, aiming to achieve upper middle-income status by 2030 (Chingarande et al., 2023). The 2021–2025 National Development Strategy (NDS 1) acknowledges the gender gap in STEM disciplines and commits to addressing this gap through inclusive and equitable access to education (Chingarande et al., 2023). Supporting this vision, the Ministry of Higher and Tertiary Education, Innovation, Science, and Technology Development developed Doctrine 5.0, a framework that outlines the significant role of universities in the industrialization process and highlights the importance of science in this development (Government of Zimbabwe, 2018). While science is recognized as a key factor for achieving Vision 2030, the vision also emphasizes gender equality as a human right and a foundation for peaceful, prosperous, and sustainable development (Chingarande et al., 2023). Consequently, inclusive science initiatives are deemed essential for Zimbabwe's industrialization agenda.

To address barriers to women's participation in higher education and STEM disciplines, the Government of Zimbabwe implemented measures in 1995, including an affirmative action admission policy, to facilitate women's entry into public universities (Government of Zimbabwe (GoZ), 2018; Chingarande et al., 2023). This policy allows women to gain admission to public higher education institutions with two points less than their male counterparts. At that time, only 25% of enrolled students were women, however by 2021, 54% of university students were female, with nearly 20% in STEM disciplines (UNESCO, 2021). The Government of Zimbabwe also provided scholarships for women enrolled in STEM disciplines (Maisiri, 2024). However, these measures have not achieved significant progress toward the desired goal, as the statistics indicate. The persistent gender disparities in higher education and STEM disciplines in Zimbabwe led to the establishment of WUA in 2002, with the specific mandate to address these disparities in education (Chingarande et al., 2023).

Networks of stakeholders interested in advancing gender inclusivity in STEM education in Zimbabwe aim to create an empowering environment for women (Isaacs et al., 2023). For example, UNICEF, along with various collaborators, calls for action from global, national, and regional actors to leverage STEM education to transform gender norms in the education system, enhance quality learning opportunities for girls, and accelerate their transition to STEM careers (Maisiri, 2024). Additionally, the Forum for African Women Educationalists (FAWE) STEM model focuses on increasing girls' participation in STEM subjects to facilitate career progression in related fields (Isaacs et al., 2023). Furthermore, there is a need to implement deliberately gendered enrollment strategies in rural schools, complemented by incentives such as STEM scholarships specifically targeting girls (Makuyana & Mbulayi, 2019).

Increasing diversity and broadening participation in STEM cannot be viewed or addressed as an independent problem; instead, it requires a holistic approach where diversity and inclusion are intentionally integrated into every aspect of the educational ecosystem (National Science Foundation, 2020). From designing learning environments. Instructional materials to training educators and implementing support interventions, every element must be carefully considered to ensure equitable opportunities for all students (Moyo, 2019). Only by embedding diversity and inclusion into the fabric of the educational system can learners of all ages and backgrounds have equitable opportunities to participate in the STEM enterprise (Moyo, 2019). As stated in chapter "Introduction", 2016, the Zimbabwean government launched the STEM Scholarship Programme. Through this initiative, the Ministry of Higher and Tertiary Education, Science and Technology Development is committed to sponsoring students willing and able to pursue STEM subjects at the A-level in public schools. The scholars noted that within the first 3 months of the program, over 3400 students had registered.

The design of curriculum and instructional materials is foundational to creating inclusive learning environments in STEM education (Moyo, 2019). STEM curricula should reflect diverse perspectives, cultures, and experiences to engage students from various backgrounds, and instructional materials should be culturally relevant, accessible, and inclusive to accommodate different learning styles and abilities (Madusise, 2019). Incorporating real-world examples and interdisciplinary approaches can help make STEM concepts more relatable and accessible to all students (Mutseekwa, 2021). Incorporating diverse perspectives, examples, and case studies into the curriculum helps students see themselves represented in STEM fields and fosters a sense of belonging. Additionally, fostering collaborative and inclusive classroom environments where all students feel valued, respected, and empowered to participate enhances learning outcomes and promotes diversity in STEM (Madusise, 2019).

In Zimbabwe, creating a supportive and inclusive classroom culture is essential for promoting diversity and participation in STEM education. Classroom environments should be welcoming, respectful, and inclusive of all students, regardless of their background or identity (Moyo, 2019). Educators can establish norms of collaboration, empathy, and mutual respect to foster a sense of belonging among students. Additionally, providing opportunities for student voice and agency empowers learners to actively engage in the learning process and contribute their unique perspectives to STEM discussions (Magwa & Mohangi, 2022).

In Zimbabwe, teachers play a central role in fostering inclusive learning environments in STEM classrooms (Svodziwa, 2021). Therefore, training and professional development programs should equip teachers with the knowledge, skills, and strategies to effectively support the diverse needs of their students (Mutseekwa, 2021). This includes training on culturally responsive teaching practices, inclusive pedagogies, and strategies for addressing implicit biases. The role of cultural and contextual factors in STEM education in Zimbabwe is essential for improving educational practices' effectiveness and inclusivity (Mashoko, 2022). In Zimbabwe, Chitera and Moyo (2021) reported that integrating cultural relevance and technological advancements boosts students' academic achievement in STEM subjects. This approach makes STEM education more accessible and relatable to African students. Understanding local challenges, linguistic diversity, and integrating culturally relevant pedagogies are crucial for enhancing the effectiveness and inclusivity of STEM education (Moyo & Ramirez, 2017). These factors are vital for developing academically sound and culturally responsive educational practices, meeting African learners' diverse needs (Chitera & Moyo, 2021). Additionally, ongoing support and mentorship can help educators continuously refine their practices to better meet the needs of diverse learners.

Interventions and support services are critical in addressing barriers to participation in STEM for underrepresented groups (Chingarande et al., 2023). These may include academic support programs, mentorship initiatives, tutoring services, and access to resources such as scholarships and financial aid. Interventions should be designed with equity in mind, ensuring that all students have access to the support they need to succeed in STEM education and careers. In Zimbabwe, the Quality Education in Science Teaching (QUEST) Project is an ongoing initiative designed to decentralise in-service training courses and empower teachers to make decisions about issues affecting them at the school level (Mukundu et al., 2017). It caters to teachers from Form 1 to Form 4. QUEST employs a cascade model of skills transfer, where trained resource persons at district resource centres train heads of science departments, who then train the teachers in their departments (Gudyanga et al., 2013). This approach provides a forum for teachers to share ideas and discuss problems, thereby enhancing the overall quality of science education. The Nziramasanga report states that 77 science district resource centres were established (Mukundu et al., 2017). Each centre is managed by two trained science advisers, who are responsible for fundraising and identifying the needs within their districts. They address these needs through dialogue and workshops, aiming to find effective solutions (Gudyanga et al., 2013).

The Science Education In-Service Teacher-Training Program (SEITT) is another staff development initiative designed for A-level science and mathematics teachers (Bhukuvhani et al., 2023). It has established resource centers managed by science advisers who hold a Diploma in Science Education from the University of Zimbabwe. SEITT provides a forum for A-level teachers to interact, share ideas, and solve problems commonly faced by isolated science teachers (Gudyanga et al., 2013; Bhukuvhani et al., 2023). Collaboration among educational institutions, community organizations, industry partners, and government agencies is essential for advancing diversity and participation in STEM in Zimbabwe (Moyo, 2019; Makuyana & Mbulayi, 2019). By working together, stakeholders can leverage their collective expertise, resources, and networks to develop innovative solutions and initiatives that address systemic barriers to inclusion. Additionally, engaging with local communities and stakeholders ensures that initiatives are culturally relevant, responsive to community needs, and sustainable in the long term (Madusise, 2019).

In Zimbabwe, implementing targeted outreach programs is essential for engaging underrepresented groups and encouraging their participation in STEM education and careers (Dube, 2018). These programs may include initiatives such as STEM summer camps, workshops, and career fairs specifically tailored to students from marginalised backgrounds. For instance, in Zimbabwe, Science camps have been organised for girls by UNESCO to provide hands-on learning experiences, exposure to role models, and information about STEM pathways, which inspire interest and confidence in pursuing STEM-related fields among girls. In 2018, the government of Zimbabwe rolled out a multimedia outreach program to increase the number of STEM students enrolling in STEM degree programs at the country's universities. There has been a noticeable increase in student enrollment in these programs. The media has played a crucial role in revolutionising the promotion of STEM careers, aligning with ZIMASSET's human capital objectives (Dube, 2018).

In Zimbabwe, mentorship plays a pivotal role in supporting the academic and professional development of individuals from underrepresented groups in STEM. Establishing mentorship programs that pair students and early-career professionals with experienced mentors from similar backgrounds provides valuable guidance, support, and networking opportunities (Nkala et al., 2024). Mentors can offer advice, share their experiences, and serve as role models, helping mentees navigate educational and career pathways in STEM and overcome barriers to success.

The international research community directs its attention towards ensuring equal opportunities in STEM education, particularly concerning gender bias and considerations related to ethnicity and socioeconomic status (Chingarande et al., 2023). Addressing bias and stereotypes within STEM fields is critical for creating an inclusive and welcoming environment for all individuals (Zvavahera, 2014). Educators, policymakers, and industry leaders must work together to challenge stereotypes, dismantle systemic barriers, and create opportunities for individuals from underrepresented groups to thrive in STEM fields (Moyo, 2019). Providing training and professional development on topics such as unconscious bias, cultural competence, and inclusive teaching practices equips stakeholders with the tools and awareness needed to create more inclusive STEM environments (AI Hamad et al., 2024).

Implementing supportive policies and practices that prioritise diversity, equity, and inclusion within educational institutions, research organisations, and industry settings is essential for promoting systemic change (Moyo, 2019). This may include initiatives such as diversity recruitment and retention efforts, inclusive hiring practices, and policies that address harassment and discrimination. By fostering a culture of inclusivity and equity, organisations can attract and retain diverse talent, drive innovation, and create more equitable opportunities for individuals from underrepresented groups in STEM (World Bank, 2020). Creating greater diversity and participation in STEM requires a comprehensive and intentional approach, considering every aspect of the educational ecosystem (Chitate, 2016). By embedding diversity and inclusion into curriculum design, educator training, classroom culture, support services, and collaborative partnerships, stakeholders can create equitable opportunities for learners of all ages and backgrounds to participate in the STEM enterprise (Pakombwele et al., 2024). Ultimately, fostering diversity and inclusion in STEM education not only benefits individual learners but also contributes to a more innovative, equitable, and sustainable society.

Conclusion

In conclusion, the trajectory of STEM education in developing countries holds immense significance for fostering sustainable development and enhancing global competitiveness in the 4IR era. As disruptive technologies reshape industries and drive global workforce demands, there is a pressing need for effective STEM education to meet evolving expectations and emerging needs. However, achieving success in STEM education requires aligning approaches with twenty-first-century skills and the demands of the 4IR era. This entails comprehensive policy reforms, investment in infrastructure and resources, fostering inclusive access to STEM education, integrating emerging technologies, and providing professional development opportunities.

In the pursuit of equitable and sustainable development, the future of STEM education in developing countries emerges as a pivotal axis around which progress orbits. Throughout this discourse, we have traversed the landscape of opportunities inherent in redefining STEM educational paradigms within these contexts. From the imperative of inclusive policy frameworks to the necessity of fostering dynamic learning environments, the path towards successful implementation requires a concerted and multifaceted approach.

As we conclude our exploration, it becomes evident that the re-alignment of STEM education approaches in developing countries is not a solitary endeavor but rather a collaborative odyssey. It necessitates the collective commitment of policymakers, educators, industry leaders, and civil society stakeholders to forge pathways that transcend traditional boundaries and empower individuals from all walks of life. By embracing innovative strategies and collaborating with stakeholders, developing countries can unlock the potential of STEM education to drive socioeconomic growth, scientific advancement, and global sustainability. Through concerted efforts and strategic alignment, the future of STEM education in developing countries can be transformed, paving the way for a brighter and more prosperous future for all.

Promoting inclusivity and diversity within STEM fields is imperative for addressing gender and socio-economic disparities, fostering innovation, and driving societal progress (Isaacs et al., 2023). Collaborative efforts among governments, educational institutions, industry partners, and community organizations are essential for implementing targeted outreach programs, mentorship initiatives, and inclusive practices that cultivate a more diverse talent pool in STEM disciplines (Mandiudza, 2021). By prioritizing inclusivity and diversity, stakeholders can create more equitable opportunities and build a future where all individuals have the opportunity to contribute to and benefit from the advancements in STEM.

At its core, the future of STEM education in developing countries embodies a vision of inclusivity, innovation, and resilience. It envisions a future where every child, regardless of their socioeconomic background or geographic location, has access to quality STEM education that nurtures their curiosity, cultivates their creativity, and equips them with the skills and mindset to thrive in an ever-evolving world.

References

AI Hamad, M. N., Adewusi, O. E., Unachukwu, C. C., Osawaru, B., & Chisom, O. N. (2024). A review on the innovative approaches to STEM education. *International Journal of Science and Research Archive., 11*(1), 244–252.

Aslam, S., Alghamdi, A. A., Abid, N., & Kumar, T. (2023). Challenges in implementing STEM education: Insights from novice STEM teachers in developing countries. *Sustainability, 15*(19), 14455. https://doi.org/10.3390/su151914455

Baingana, J. K. (2024). Enhancing STEM education in Africa through collaborative learning and technological integration. *Research Invention Journal of Research in Education, 3*(3), 68–76.

Bhukuvhani, C., Nezamndonyi, J., & Mutanga, P. (2023). Advancement of Science, Technology and Innovation (STI) education in Zimbabwe: Policies and practices. In B. Chirinda, L. Sibanda, J. Vere, & G. Sunzuma (Eds.), *Science, mathematics and technology education in Zimbabwe* (pp. 27–44). Peter Lang. issn:2576-358x (print), issn:3598 (online).

Botha, Z. (2022). Presidential address: Collaboration and diversity in STEM. *Journal of the Southern African Institute of Mining and Metallurgy, 122*(9), 487–496. https://doi.org/10.17159/2411-9717/2022

Chara, T. (2016). STEM: The merits, demerits. *The Sunday Mail.* Retrieved from https://www.Sundaymail.co.zw/stem-the-merits-demerits

Chauraya, M. (2023). Education for sustainable development and STEM education: Implications for an interdisciplinary approach to mathematics education in Zimbabwe development and STEM education. In B. Chirinda, L. Sibanda, J. Vere, & G. Sunzuma (Eds.), *Science, mathematics and technology education in Zimbabwe*. Peter Lang.

Chimhenga, S. (2017). Transforming digital learning and teaching in institutions of higher learning in Zimbabwe: The case of Zimbabwe Open University, Bulawayo Regional Campus. *Global Journal of Advanced Research, 4*(3), 121–129.

Chingarande, S. D., Muchabaiwa, W., & Muyambo, T. (2023). *Bridging the gender equality gap in science at the women's university in Africa, Zimbabwe*. Final Technical Report submitted to IDRC.

Chirume, E., & Kwabena, D. (2016). Toward a better understanding of the access to education through online learning in the case of Zimbabwe. *International Journal of Advanced Research, 4*(7), 1304–1313. https://doi.org/10.21474/ijar01/1031

Chisom, O. N., Unachukwu, C. C., & Osawaru, B. (2024). STEM education advancements in African contexts: A comprehensive review. *World Journal of Advanced Research and Reviews, 21*(1), 145–160.

Chitanana, L. (2015). The integration of instructional technology by teacher educators at a State University in Zimbabwe: Are they leading by example ? *Zimbabwe Journal of Educational Research, 27*(1), 1–26.

Chitate, H. (2016). Science, Technology, Engineering and Mathematics (STEM): A case study of Zimbabwe's educational approach to industrialisation. *World Journal of Education, 6*, 27–35.

Chitera, J., & Moyo, P. V. (2021). Integrating cultural knowledge in science teaching at ordinary level in Bulawayo, Zimbabwe. *Transatlantic Journal of Multidisciplinary Research, 3*(1), 1–23. https://doi.org/10.5281/zenodo.4724222

Dodo, O., Makwerere, D., & Matavire & Parwada, C. (2013). An analysis of challenges faced by students learning in virtual and open distance learning system: A case of Bindura University of Science Education (BUSE). *Educational Research and Reviews, 8*(11), 700–707.

Dube, P. S. (2018). An investigation of the Science Technology Engineering and Mathematics (STEM) initiatives in the Zimbabwean education system. *International Journal for Cross-Disciplinary Subjects in Education (IJCDSE), 9*(1), 3329–3334.

Dube, T., Ncube, S. B., Mapuvire, C. C., Ndlovu, S., Ncube, C., & Mlotshwa, S. (2021). Interventions to reduce the exclusion of children with disabilities from education: A Zimbabwean perspective from the field. *Cogent Social Societies, 7*(1), 1–18. https://doi.org/10.1080/23311886.2021.1913848

Dzinoreva, T., & Mavunga, G. (2022). Integrating ICTs into the Zimbabwean secondary school pre-service teachers' curriculum. *Journal of Education, 88*, 53–68.

Dzinotyiweyi, M., & Taddese, A. (2020). *EdTech in Zimbabwe: A Rapid Scan*. EdTech Hub Country Scan. https://doi.org/10.5281/zenodo.3903838. Retrieved from https://docs.edtechhub.org/lib/G4UUX5P3

Gadzirayi, C. T., Bongo, P. P., Ruyimbe, B., Bhukuvhani, C., & Mucheri, T. (2016). *Diagnostic study on the status of STEM in Zimbabwe*. Bindura University of Science Education and Higher Life Foundation.

Gambanga, N. (2016). STEM Students at Public Schools in Zimbabwe Offered Free Education. *Techzim*.

Gomba, C. (2016). Transforming rural secondary schools in Zimbabwe through technology: Lived experiences of student computer users. *International Online Journal of Education and Teaching (IOJET), 3*(2), 108–120.

Government of Zimbabwe. (2018). *Doctrine education 5.0 heritage, innovation, industrialisation. The modernisation & industrialisation of Zimbabwe through science education & development*. Ministry of Higher and Tertiary Education, Science and Technological Development [MoHTEISTD].

Gudyanga, A., Gudyanga, E., & Mutemeri, J. (2013). Innovative science trends that have occurred in Zimbabwe. *Asian Social Science, 9*(12), 50–56. https://doi.org/10.5539/ass.v9n11p50

Gweshe, G. T., & Chiware, M. (2023). Gender-inclusive education in Science, Technology, Engineering, and Mathematics (STEM) fields in postindependence Zimbabwe. In D. Mhlanga & E. Ndhlovu (Eds.), *Economic inclusion in post-independence Africa. Advances in African economic, social and political development*. Springer. https://doi.org/10.1007/978-3-031-31431-5_16

Isaacs, N., Middleton, L., Essop, R., Lynch, I., Fluks, L., Chikwanha, M., Chinyemba, F., Gumede, Z., Djoukouo, F., Kuetche, I., Maseko, N., Ndinda, C., Tamangani, R., Tendenedzai, J., Agugua, A., & Van Rooyen, H. (2023). *Strengthening gender equality and inclusivity in the national system of Science, Technology and Innovation: Zimbabwe country profile*. HSRC.

Kalolo, J. F. (2016). Re-aligning approaches for successful implementation of STEM education in today's elementary schools in developing countries: Policy commitments and practices. *Journal of Education and Literature, 4*(2), 61–76.

Kangara, C. T. C., Gocha, M., Tsokota, T., & Marovah, T. (2022). Assessing readiness for teaching and learning using ICTs in Zimbabwean Secondary Schools in Gweru Distrct, Zimbabwe. *Creative Education, 14*(5). https://doi.org/10.21203/rs.3.rs-1652431/v1

Kumba, H., Olanrewaju, O., Akpan, J., & Mafuratidze, F. (2024). *Engineering education and industry synergy in Zimbabwe: A divide and redirection*. IntechOpen.

Mabhanda, W. (2016). Opportunities and factors affecting adoption of STEM education: The case of Gweru Polytechnic first year commerce students. *International Journal of Business Marketing and Management, 1*(5), 1–8.

Madusise, S. (2019). Possibilities for implementation of STEM education in Zimbabwean under-resourced mathematics classrooms. *Journal of African Interdisciplinary Studies, 3*(7), 110–124.

Magwa, L., & Mohangi, K. (2022). Using theoretical frameworks to analyze democratic student–teacher engagement and autonomous learning for academic achievement in Zimbabwe. *Frontiers of Education, 7*. https://doi.org/10.3389/feduc.2022.925478

Maisiri, J. (2024). Preparing Zimbabwean women for the future of work: The Role of STEM education in the so-called fourth industrial revolution. *South African Review of Sociology, 54*(1), 5–22. https://doi.org/10.1080/21528586.2024.2311399

Makuyana, A., & Mbulayi, S. P. (2019, July 8–11). *Beyond the enclave: Expediting the challenges and prospects of young girls doing STEM subjects in public high schools in rural Zimbabwe*. Being a paper presented at Al-Azhar International Conference Center Cairo, Egypt by Conference Of Rectors, Vice Chancellors & Presidents of Association African Universities. Theme: Role of Higher Education Institutions in Promoting the Continental Education Strategy for Africa (CESA 16–25).

Mandiudza, M. (2021). Factors motivating female students from masvingo, zimbabwe—The choice of taking natural sciences at university. *European Journal of Social Sciences Studies, 6*(2). https://doi.org/10.46827/ejsss.v6i2.1001

Maphosa, V., Dube, B., & Jita, T. (2021). Sustainable information and learning access at a rural university in Zimbabwe through a mobile application. *International Journal of Information and Education Technology, 11*(2), 82–87. https://doi.org/10.18178/ijiet.2021.11.2.1493

Mashoko, D. (2022). Indigenous artefacts and physics curriculum: Teaching science as a cultural way of knowing. *Culture Studies of Science Education, 17*, 863–874. https://doi.org/10.1007/s11422-021-10085-1

Maune, A. (2023). Adoption and use of eLearning platforms by universities in developing countries: Evidence from Zimbabwe. *Cogent Education, 10*(2). https://doi.org/10.1080/2331186X.2023.2287905

Mhlanga, D. (2024). Digital transformation of education, the limitations and prospects of introducing the fourth industrial revolution asynchronous online learning in emerging markets. *Discovery Education, 3*(32). https://doi.org/10.1007/s44217-024-00115-9

Ministry of Primary and Secondary Education. (2014). *Curriculum framework for primary and secondary education 2015–2022*. Zimbabwe Ministry of Primary and Secondary Education.

Ministry of Primary and Secondary Education. (2015a). *Curriculum framework for primary and secondary education 2015–2022*. Zimbabwe Ministry of Primary and Secondary Education.

Ministry of Primary and Secondary Education. (2015b). *Mathematics syllabus forms 1–4, 2015–2022*. Zimbabwe Ministry of Primary and Secondary Education.

Moyo, M. (2019). Towards implementing a STEMitised curriculum in inclusive secondary school contexts: The case of five selected schools in Mzilikazi District of Bulawayo. *Zimbabwe Journal of Educational Research, 31*(3), 470–484.

Moyo, M., & Ndlovu, C. (2023). Situating environmental science education to reflect the pedagogy of place and the socio-cultural context in Zimbabwe's higher education institutions curricular. In B. Chirinda, L. Sibanda, J. Vere, & G. Sunzuma (Eds.), *Science, mathematics and technology education in Zimbabwe: Research, policy and practice* (pp. 159–189). Peter Lang.

Moyo, P. V., & Berdut, I. R. (2017). Interrogating the implementation of STEM education in Zimbabwe. *The International Journal of Humanities and Social Studies, 5*(11), 332–337.

Moyo, P. V., & Ramirez, I. (2017). Inclusion through multicultural science education. *The International Journal of Humanities and Social Studies, 6*(11), 326–331.

Mpofu, V. (2019). A theoretical framework for implementing STEM education. In K. G. Fomunyam (Ed.), *Theorizing STEM education in the 21st century*. IntechOpen. isbn:978-1-78985-702-3.

Mpofu, V., & Mutseekwa, C. (2022). Online stem teaching of practical chemistry: Challenges and possibilities. *Academic Voices*, 115–127. https://doi.org/10.1016/b978-0-323-91185-6.00014-8

Mukundu, K. C., Chineka, R., & Madzudzo, A. (2017). The framing and reframing of science education, training and research in Zimbabwe: Past, present and future. In *The world of science education*. Brill. Retrieved March 1, 2025, from https://brill.com/view/book/edcoll/9789463510899/BP000010.xml

Mutseekwa, C. (2021). STEM practices in science teacher education curriculum: Perspectives from two secondary school teachers' colleges in Zimbabwe. *Journal of Research in Science, Mathematics and Technology Education, 4*, 75–92. https://doi.org/10.31756/jrsmte.422

Mutsvangwa, A., & Zezekwa, N. (2021). STEM education: A ray of hope for African countries. *Unnes Science Education Journal, 10*(2), 79–89. https://doi.org/10.15294/usej.v10i2.45746

Mwembe, D., Zulu, I. L., & Mdlongwa, P. (2022). Needs assessment for facilitators in teaching of STEM subjects in under resourced rural schools: Case of Lupane and Hwange districts in Zimbabwe. *International Journal of Professional Development, Students and Learning, 4*(1), ep2202. https://doi.org/10.30935/ijpdll/11666

Nadelson, L. S., & Seifert, A. L. (2017). Integrated STEM defined: Contexts, challenges, and the future. *The Journal of Educational Research, 110*(3), 221–223. https://doi.org/10.1080/00220671.2017.1289775

National Science Foundation. (2020). *STEM education for the future: 2020 visioning report*. Retrieved from https://www.nsf.gov/ehr/Materials/STEM%20Education%20for%20the%20Future%20-%202020%20Visioning%20Report.pdf

Nezandonyi, J. (2023). Embedding education for sustainable development in the teaching of technology courses in Zimbabwe: Policy and practice. In B. Chirinda, L. Sibanda, J. Vere, & G. Sunzuma (Eds.), *Science, mathematics and technology education in Zimbabwe: Research, policy and practice* (pp. 159–189). Peter Lang.

Nkala, B., Sibanda, V. M., Ndhlovu, J., Hobane, L., & Singh, S. (2024). Challenges and drivers of industrialist propensity among chemical engineering students in STEM institution in Zimbabwe: Towards a conceptual framework. *International Journal of Environment, Engineering and Education, 6*(1), 26–37. https://doi.org/10.55151/ijeedu.v6i1.126

Parawira, W. (2016). Science in our natural world. *The Sunday Mail*. Retrieved from http://www.sundaymail.co.zw

Pakombwele, A., Hlatshwayo, I., Sikhangele, M., Gumbo, R., & Ncube, S. (2024). Quality and inclusive education in Zimbabwean rural early childhood development centres. *International Journal of Research and Innovation in Social Science (IJRISS), 8*(1), 1242–1253.

Ruparanganda, F., Mukundu, C., & Munjoma, G. (2012). Students perceptions on the effectiveness and advantages of the open distance electronic learning methodology (ODeL) over the traditional face to face learning methodology, "two years on.". *Zimbabwe Journal of Educational Research, 24*(3), 291–302.

Shava, G. N. (2021). Quality education for sustainable development in Zimbabwean higher education. In K. S. Adeyemo (Ed.), *The education systems of Africa. Global education systems*. Springer. https://doi.org/10.1007/978-3-030-44217-0_6

Shizha, E., & Kariwo, M. T. (2011). *Education and development in Zimbabwe: A social, political, and economic analysis*. Sense Publishers.

Svodziwa, M. (2021). *STEM education as a human capital development strategy for the industrialization and modernization of Zimbabwe*. Retrieved from SSRN: https://ssrn.com/abstract=3788852 or https://doi.org/10.2139/ssrn.3788852

UNESCO. (2021). UNESCO science report. In *The race against time for smarter development* (pp. 1–758). UNESCO Publishing.

World Bank. (2020). *Revitalizing Zimbabwe's tertiary education sector to support a robust economic recovery*. The International Bank for Reconstruction and Development, World Bank.

Yingi, E., Hlungwani, P. M., & Nyagadza, B. (2022). The Fourth Industrial Revolution (4IR) in the heart of the SDG agenda: The role of education in Zimbabwe. *Africa Review, 14*(2), 213–229. https://doi.org/10.1163/09744061-01402001

Zhou, M. (2017). *STEM implementation in rural secondary schools: A case of Kwekwe Zhombe district*. Undergraduate research project. Midlands State University.

Zimbabwe Flying Labs. (2020). *In Zimbabwe, a new STEM education program gives wings to young learners*. Retrieved from https://blog.flyinglabs.org/2021/02/08/in-zimbabwe-a-new-stem-education-program-gives-wings-to-young-learners/

Zvavahera, P. (2014). The presidential scholarship programme in Zimbabwe: A living case of the political will in promoting regionalisation in higher education. *Journal of Case Studies in Education, 6*, 1–9.

Looking Ahead

Abstract This chapter examines the future STEM education within the socio-economic context of poverty in Zimbabwe, emphasizing the interplay between socio-economic factors and access to quality education. Effective STEM education has the potential to act as a transformative force for individual empowerment and economic development. The chapter proposes several strategies to improve STEM education, including adopting innovative, learner-centered pedagogies and providing ongoing professional development for educators. It recommends policy interventions prioritizing investments in STEM infrastructure and creating curricula that respond to national and regional economic needs. Furthermore, the chapter underscores the necessity of fostering awareness and interest in STEM disciplines among students, recommending engaging and interactive teaching methodologies to ignite curiosity and creativity. It also highlights the importance of introducing more STEM-related programs within educational institutions and promoting extracurricular activities such as science clubs, which facilitate hands-on learning experiences. Ultimately, the chapter posits that STEM education should be regarded not merely as an educational reform but as a crucial mechanism for achieving social and economic empowerment in impoverished contexts.

This book uses Zimbabwe as an example to explore STEM education in the context of poverty. Studying the intersection of poverty and STEM education in Zimbabwe helps to better understand how socioeconomic factors shape educational access and quality. As stated by Jones (1992):

> We are where we live…. Our immediate surroundings determine the nature of the schooling our children receive, the kinds of jobs we get, the cost of our auto insurance, the quality of health care and police protection available to us, the merchandise we find at local stores and the prices we pay for it. Our sense of ourselves as individuals and family members is reflected in the tranquillity of tree-lined streets or the filthy clutter of vacant lots. (p. 20)

Our detailed exploration of STEM education in the context of Zimbabwe's poverty revealed to us challenges and opportunities for meaningful change. It underscored the importance of equitable education as a pathway to empowerment and economic development. During the compilation of this book, we realised that STEM education has the potential to be a transformative force in alleviating poverty, fostering

B. Chirinda et al., *STEM Education in Poverty*,
https://doi.org/10.1007/978-3-031-92490-3_10

economic growth, and bridging the global knowledge divide. However, for many developing countries such as Zimbabwe, systemic challenges—such as inadequate infrastructure, a lack of qualified teachers, and limited access to technology—continue to hinder progress. Addressing these challenges requires a multi-faceted approach involving governments, non-governmental organizations (NGOs), educational institutions, and industry partnerships. Below, we present strategies to enhance STEM education in developing countries, using Zimbabwe as an example.

General Strategies to Enhance STEM Education in Zimbabwe

Effective STEM pedagogy is hampered by an education system that emphasises the curriculum and teacher-centred focus instead of focusing on how students can be taught or learn. The empirical findings from chapter "STEM Education Pedagogy in Developing Country Contexts: Relevance and Challenges" revealed that in Zimbabwe, STEM teachers must apply inventive pedagogies and suitable learner-centred STEM teaching approaches. Teachers are the key agents of STEM education, and they need to be well-equipped with the knowledge, skills, and attitudes necessary to teach STEM effectively. They should receive quality pre-service and in-service training to enhance their content knowledge, pedagogical skills, and confidence in STEM subjects. Teachers should also have access to continuous professional development opportunities that can update their knowledge and skills, as well as expose them to innovative best practices in STEM education. Additionally, they should be supported by a network of peers, experts, and resources that can provide guidance, feedback, and collaboration.

While a lack of resources and other challenges hinder effectiveness, opportunity lies in promising pathways of policy and practice. Policy interventions must prioritize investment in STEM infrastructure and the development of localized curricula that align with Zimbabwe's national and regional economic priorities. Successful examples from countries like India, Kenya, and Brazil show that STEM-focused educational reforms—such as coding boot camps, maker spaces, and vocational training programs—can empower youth with the skills needed for employment and entrepreneurship (Muralidharan & Singh, 2020). Ultimately, STEM education in poverty-stricken regions should not be seen merely as an educational reform but as a fundamental tool for social and economic empowerment. In Zimbabwe, there is a varying degree of success in the implemented competence-based curriculum, emphasising inquiry-based learning and STEM education. In this regard, the curricula for STEM subjects in developing countries should be updated to reflect the current and future needs of society and the economy. The curricula should also include clear and relevant learning outcomes for each topic and use various assessment methods to measure students' understanding and application of STEM concepts. The curricula should also promote the integration of STEM subjects, rather than teaching them in isolation, to show the connections and interdisciplinary nature of STEM fields.

There is a need for more STEM-related programs and courses in Zimbabwean schools and universities to improve the opportunities for students to pursue careers in these fields. Another challenge is the need for increased awareness about the opportunities and importance of STEM education and interest in STEM subjects among students. This can be attributed to a lack of exposure and understanding of the relevance and importance of these fields in today's world. Students should be motivated and inspired to pursue STEM education and careers by exposing them early to the benefits and opportunities of STEM fields. This can be done using engaging and interactive teaching methods, such as demonstrations, experiments, projects and games that can spark students' curiosity and creativity. Students should also have access to extracurricular activities such as science clubs, which provide hands-on and collaborative learning experiences. Moreover, students should be able to interact with role models and mentors from STEM professions who can share their stories and advice and challenge stereotypes and biases that may discourage students, especially girls, from STEM.

It is important to ensure that Zimbabwean schools, particularly those in underserved areas, have the necessary infrastructure for reliable internet connectivity and access to digital devices (Chirume & Kwabena, 2016). The World Bank (2019) emphasizes that digital literacy and STEM competencies are essential for nations transitioning from resource-based to knowledge-based economies. Developing countries risk being left behind in the global digital revolution without these competencies. Technology and digital learning offer a promising pathway for expanding access to STEM education. Mobile learning, open educational resources (OER), and initiatives like One Laptop per Child (OLPC) have demonstrated that cost-effective digital tools can enhance learning outcomes even in resource-constrained environments (Trucano, 2016). However, as Czerniewicz and Brown (2014) highlight, digital interventions must be contextually adapted to local infrastructure and pedagogical needs to ensure sustainability. Making digital resources and tools accessible to students with disabilities and ensuring that the benefits of digital literacy are extended to all learners is also crucial (Dube et al., 2021). Implementing measures to make digital devices and internet access more affordable and reducing financial barriers that may hinder access for economically disadvantaged students is important (Chirume & Kwabena, 2016).

Furthermore, there is a need to establish community-based technology centres where students can access digital resources outside of school hours (Kalolo, 2016). Community Information Centres (CICs) in Zimbabwe are facilities located in rural and marginalized communities, equipped with digital technologies such as computers, internet access, printing and photocopying, faxing, scanning, laminating, and gaming services. These centres serve as additional learning, collaboration, and skill development spaces. In Zimbabwe, digital literacy awareness campaigns are important for raising awareness about the importance of digital literacy and the resources available to students. These campaigns can involve parents, communities, and local businesses to create a supportive ecosystem for digital skill development (Mabhanda, 2016).

There is a need for collaborative effort among policymakers, educators, and the private sector to create inclusive and sustainable STEM ecosystems that enable young people to break the cycle of poverty and contribute meaningfully to their societies. In addition, the Zimbabwean government should establish partnerships with countries excelling in STEM. These collaborations should focus on rural schools, prioritising the development of necessary infrastructure and curriculum enhancements (Makuyana & Mbulayi, 2019). Developing countries like Zimbabwe can learn from and tap into the successes of other developing nations, such as Singapore, South Korea, Malaysia, and Taiwan, which have implemented effective STEM initiatives (Ngwenya et al., 2017).

Conclusion

As we finalize writing this book, we reflect on how the process has been a rewarding experience. By leveraging emerging technologies, fostering equitable access, and investing in teacher development, Zimbabwe and other developing nations can harness the full potential of STEM education to drive innovation and development in the twenty-first century. Research indicates that investment in STEM education leads to long-term economic growth. Hanushek and Woessmann (2015) argue that cognitive skills correlate strongly with national economic performance, particularly in science and mathematics. Moreover, equitable access to STEM education is crucial. Gender disparities, rural-urban divides, and socio-economic inequalities persist in many low-income regions. According to UNESCO (2021), girls and marginalised groups in developing countries encounter significant barriers to STEM participation due to cultural norms, lack of role models, and inadequate policy interventions. Strategies such as community-driven STEM programs, mentorship initiatives, and teacher capacity-building programs have demonstrated promise in bridging these gaps (Boeren, 2020). Concluding this book on a positive note, it is important to highlight that the Zimbabwean government has prioritised access to education, particularly for marginalised groups such as girls and children from low-income backgrounds. By implementing inclusive policies and programmes, efforts have been made to increase enrollment rates and reduce the gender gap. For instance, the government has provided financial aid and scholarships to students from disadvantaged backgrounds and initiated programs to promote girls' education and reduce child marriage rates.

References

Boeren, E. (2020). *Lifelong learning and development: A perspective from the global south.* Routledge.

Chirume, E., & Kwabena, D. (2016). Toward a better understanding of the access to education through online learning in the case of Zimbabwe. *International Journal of Advanced Research, 4*(7), 1304–1313. https://doi.org/10.21474/ijar01/1031

Czerniewicz, L., & Brown, C. (2014). The habitus of digital 'strangers' in higher education. *British Journal of Educational Technology, 45*(1), 44–53.

Dube, T., Ncube, S. B., Mapuvire, C. C., Ndlovu, S., Ncube, C., & Mlotshwa, S. (2021). Interventions to reduce the exclusion of children with disabilities from education: A Zimbabwean perspective from the field. *Cogent Social Societies, 7*(1), 1–18.

Hanushek, E. A., & Woessmann, L. (2015). *The knowledge capital of nations: Education and the economics of growth.* MIT Press.

Jones, J. (1992). *The dispossessed.* Basic Books.

Kalolo, J. F. (2016). Re-aligning approaches for successful implementation of STEM education in today's elementary schools in developing countries: Policy commitments and practices. *Journal of Education and Literature, Research Academy of Social Sciences, 4*(2), 61–76.

Mabhanda, W. (2016). Opportunities and factors affecting adoption of STEM education: The case of Gweru Polytechnic first year commerce students. *International Journal of Business Marketing and Management, 1*(5), 1–8.

Makuyana, A., & Mbulayi, S. P. (2019, July 8–11). *Beyond the enclave: Expediting the challenges and prospects of young girls doing STEM subjects in public high schools in rural Zimbabwe.* Paper presented at Conference of Rectors, Vice Chancellors & Presidents of Association African Universities.

Muralidharan, K., & Singh, A. (2020). Improving school-to-work transitions through STEM education in India. *Journal of Development Economics, 146*, 102520.

Ngwenya, B., Pelser, T., & Sibanda, R. (2017). Success as the knowledge economy the success or failure of Zimbabwe's STEM initiative. In *International conference on intellectual capital and knowledge management and organisational learning* (pp. 169–178). Academic Conferences International.

Trucano, M. (2016). *Education and technology: Policy, planning, and practice in developing countries.* World Bank.

UNESCO. (2021). *Cracking the code: Girls' and women's education in STEM.* UNESCO Publishing.

World Bank. (2019). *The changing nature of work: World development report 2019.* World Bank.

The manufacturer's authorised representative in the EU is Springer
Nature Customer Service Centre GmbH, Europaplatz 3, 69115 Heidelberg,
Germany. If you have any concerns regarding our products, please
contact ProductSafety@springernature.com

Printed and bound by CPI Group (UK) Ltd, Croydon, CR0 4YY
01/10/2026
02225977-0018